Physical Fundamentals
of
Materials Science

D. S. Tulloch, M.Sc., A.Inst.P., M.I.E.E.
(Senior Lecturer, Twickenham College of Technology)

LONDON
BUTTERWORTHS

THE BUTTERWORTH GROUP

ENGLAND
Butterworth & Co (Publishers) Ltd
London: 88 Kingsway WC2B 6AB

AUSTRALIA
Butterworth & Co (Australia) Ltd
Sydney: 20 Loftus Street
Melbourne: 343 Little Collins Street
Brisbane: 240 Queen Street

CANADA
Butterworth & Co (Canada) Ltd
Toronto: 14 Curity Avenue, 374

NEW ZEALAND
Butterworth & Co (New Zealand) Ltd
Wellington: 49/51 Ballance Street
Auckland: 35 High Street

SOUTH AFRICA
Butterworth & Co (South Africa) (Pty) Ltd
Durban: 33/35 Beach Grove

First published 1971

© The Author 1971

ISBN 0 408 70096 3 (Standard)
      0 408 70097 1 (Limp)

Printed in Hungary

PHYSICAL FUNDAMENTALS
OF
MATERIALS SCIENCE

Preface

This concise introductory text is for students of electrical, mechanical and civil engineering who have taken ONC or Advanced Level GCE and are reading for HNC, HND, C.Eng. and first Degree qualifications.

It is hoped that the treatment will encourage a balanced course in which numerical work is plentiful but tempered with a proportion of descriptive work. There is scope for the student to carry out much experimental verification in the teaching laboratory. Such verifications are a vital part of any course and could include the following.

The determination of e and e/m_e. Determination of the Rydberg constant. Determination of a radioactive half-life. A study of radiation absorption, and practice in the use of counters, ratemeters and scalers. A study of x-ray equipment, x-radiation and x-ray crystallographic methods. The determination of a crystallographic lattice constant by electron diffraction. Measurement of temperature coefficients of resistance for metals and intrinsic semi-conductors. Observation of the experimental evidence for electron and hole conduction in samples of 'n' and of 'p' type semi-conductor.
Naturally the list should be greatly extended.

The orbital approach to extra-nuclear structure has not been used despite its current popularity in elementary courses. Use of the Rutherford–Bohr planetary model, in spite of its limitations, is felt to be preferable at this level, for it can provide an interesting standard of numerical work needing only relatively simple mathematics. The orbital approach requires a great deal to be taken on trust until a reasonably full treatment involving the derivation, interpretation and solutions of Schroedinger's wave equations can properly be given.

Throughout the book the SI (International System of Units) is preferred.

D. S. TULLOCH

Contents

Contents

Contents

Contents

Chapter 1

Revision of Essentials

1.1. GENERAL PHYSICS

1.1.1. Introduction

This chapter is in no way meant to provide a preliminary course in physics but is intended to help the student to recall some of the relevant physical concepts, and the derivation of certain relationships, which are used in the following chapters.

Throughout the book SI* units are adhered to except that electron volts and atomic mass units are used in atomic and nuclear discussions, but SI equivalents are always quoted. In this coherent system there are seven, independent, basic physical quantities and two supplementary dimensionless quantities. From these *all* other physico-chemical quantities are derived. The seven basic, and two supplementary, quantities are listed below.

Quantity	Name	Unit symbol
Length (l)	metre	m
Time (t)	second	s
Mass (m)	kilogramme	kg

* Système International d'Unités—recently generally agreed, and adopted by ISO (International Organisation for Standardisation) of which the· BSI is a Member Body, the IUPAP, the IUPAC and the International General Conference on Weights and Measures, amongst others.

Quantity	Name	Unit symbol
Electric current (I)	ampere	A
Thermodynamic temperature (T)	kelvin	K
Quantity of substance (n)	mole	mol
Luminous intensity (I_v)	candela	cd

The two supplementary quantities are:

Plane angle	radian	rad
Solid angle	steradian	sr

Full details of all these units are given in Sub-section 1.1.4.

1.1.2. Kinematic Space–Time Relationships

The subject of kinematics only involves the first two of the above seven basic physical quantities, and the relationships which are derived here are widely used.

Let s represent distance, v represent velocity,
t represent time, a represent acceleration.

By definition
$$\frac{dv}{dt} = a$$

$$dv = a\,dt$$

if the velocity is u at time 0 and is v at time t

$$\int_u^v dv = \int_0^t a\,dt$$

whence
$$v - u = at$$

$$v = u + at \qquad \ldots(1.1)$$

By definition
$$\frac{ds}{dt} = v$$

$$ds = v\,dt$$

$$= (u + at)\,dt \qquad \text{[from (1.1)]}$$

2

If the distance is 0 at time 0 and is s at time t

$$\int_0^s ds = \int_0^t (u+at) \, dt$$

whence $\qquad\qquad s = ut + \tfrac{1}{2}at^2 \qquad\qquad \ldots(1.2)$

Eliminating t between Equations (1.1) and (1.2) provides a third relationship:

$$v^2 = u^2 + 2as \qquad\qquad \ldots(1.3)$$

1.1.3. Particle Dynamics

This subject takes account of the first three of the seven basic physical quantities. There are three essential laws involved in elementary dynamics, first formulated by Newton and known as Newton's laws of motion. They are as follows:

1. All bodies continue in their state of rest, or of uniform motion in a straight line, unless compelled to change that state by external forces.
2. The rate of change of momentum of a body is proportional to the force impressed on the body, and the change of momentum is in the direction of action of that force.
3. To every action there is an equal and opposite reaction.

Consider the case of a particle of mass m travelling with a constant speed v in a circular orbit of radius R and centre O. The momentum of the particle is a vector quantity $\overrightarrow{mv}$, always tangential to the orbit and therefore, although of constant magnitude, has a constantly changing direction. By Newton's first law there is a force acting on the particle; let this force be $\vec{F}$.

In *Figure 1.1 (i)*, let the particle be at position (1) at time t and at position (2) at time $t+dt$, having moved through the angle $d\theta$ in time dt, then

$$\frac{d\theta}{dt} = \frac{v}{R} \qquad\qquad \ldots(1.4)$$

Figure 1.1(ii) is a vector diagram showing the momenta at (1) and (2) as well as the change of momentum $\overrightarrow{d(mv)}$ which occurs in time dt and which is necessary to obtain the momentum at (2) from that at (1). From the geometry of this vector diagram,

$$\left|\overrightarrow{d(mv)}\right|^* = mv \cdot d\theta \qquad\qquad \ldots(1.5)$$

* $\left|\overrightarrow{d(mv)}\right|$ means the *magnitude* of the vector $\overrightarrow{d(mv)}$

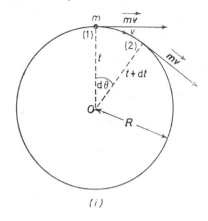

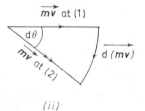

Figure 1.1. Circular motion; (i) general, (ii) momentum vector diagram

Assuming a system of units which makes its constant of proportionality unity, Newton's second law gives

$$\vec{F} = \frac{\overrightarrow{d(mv)}}{dt} \qquad \ldots(1.6)$$

hence $$|\vec{F}| = \overrightarrow{mv} \cdot \frac{d\theta}{dt} \qquad \text{[from (1.5)]}$$

and $$|\vec{F}| = \frac{mv^2}{R} \qquad \text{[from (1.4)]}$$

$\vec{F}$ is in the direction of $\overrightarrow{d(mv)}$ (from Newton's second law), that is, the force is always perpendicular to the momentum vector. It is thus always towards O, the centre of motion, and is therefore

4

termed *centripetal*. $\dfrac{mv^2}{R}$ is the centripetal force necessary to enable the particle of mass m to travel in the prescribed path.

1.1.4. Units and Dimensions

So far it has been sufficient to rely upon the intuitive concept of length, time and mass. Now the precise physical significance of these, and the other, basic quantities are considered. As indicated in Sub-section 1.1.1 there are seven basic physical quantities and two supplementary quantities in the SI, and these are defined by the following standards.

The metre—previously defined as the distance between two engraved lines on a platinum-iridium alloy bar. Now defined as 1 650 763.73 wavelengths, in vacuum, of the radiation corresponding to the transition between the levels $2p_{10}$ and $5d_5$ of the krypton-86 atom.

The second—previously defined as $\dfrac{1}{86\,400}$ th. part of the mean solar day. Now defined as the duration of 9 192 631 770 periods of the radiation corresponding to the transition between the two hyperfine levels of the ground state of the caesium-133 atom.

The kilogramme—the mass of the international prototype comprising a platinum-iridium alloy mass kept at the International Bureau of Weights and Measures (BIPM) at Sèvres near Paris.

The ampere—That electric current which, if maintained in each of two straight parallel conductors of infinite length, of negligible circular cross section, and placed one metre apart in a vacuum, would produce between these conductors a force equal to 2×10^{-7} newton per metre of their length.

The kelvin—formerly the 'degree kelvin', is the fraction $\dfrac{1}{273.16}$ of the thermodynamic temperature of the triple point of water.

The mole—the amount of substance which contains the same number of elementary units as there are atoms in 0.012 kilogramme of the isotope carbon-12. The elementary units must be specified and may be, for example, atoms, molecules, ions etc., or groups of these.

The candela—the luminous intensity normal to a surface of $\dfrac{1}{600\,000}$ square metre of a 'Black Body' at the temperature of freezing platinum under a pressure of 101 325 newton per square metre (one standard atmosphere).

The radian—that plane angle subtended at the centre of a circle by an arc of the circle equal in length to the radius of the circle.

The steradian—that solid angle subtended at the centre of a sphere which encloses a surface on the sphere of area equal to the square of the radius.

All the derived quantities of the physical, chemical and engineering sciences are based upon the above fundamental quantities. Of the derived quantities the following few are of general importance. They involve only length, time and mass, which can be symbolised by l, t and m (referring to the physical quantities and not to their units of measurement).

Area—defined as quantity of surface, symbolised by l^2 and having units m^2.

Volume—defined as quantity of three dimensional space, symbolised by l^3 and having units m^3.

Velocity—defined as rate of change of position, symbolised by $\dfrac{l}{t}$ and having units m s^{-1}.

Acceleration—defined as rate of change of velocity, symbolised by $\dfrac{l}{t^2}$ and having units m s^{-2}.

Force—defined by Newton's second law of motion as that quantity which induces an acceleration in a constant mass, symbolised by $m\dfrac{l}{t^2}$ and having units kg m s^{-2} which are given the special name newtons (unit symbol N).

Work and *Energy*—the former is defined as the product of a force and the distance through which it moves its point of application in the direction of the force. Energy is that quantity which enables a system to perform work, and has many forms including kinetic (energy of motion) and potential (energy of position). These are both symbolised by $m\dfrac{l}{t^2}$ and have units kg m^2 s^{-2} which are also newton metres (N m) and these are given the special name joules (unit symbol J).

Power—defined as rate of doing work, or as rate of energy conversion. It is symbolised by $m\dfrac{l^2}{t^3}$ and has units kg m^2 s^{-3} which are also newton metres per second (N m s^{-1}), which in turn are joules per second (J s^{-1}) and these are given the special name watts (unit symbol W).

Those expressions involving such symbols as l, t and m, for example $m\dfrac{l}{t^2}$, are termed the *dimensions* of the quantity concerned (work, or energy, in the example); and in a calculation *dimensional analysis* is of great value in checking the validity of the calculation procedure.

Many other derived quantities are involved in the various branches of pure and applied science, and some of those relating to electricity and magnetism are dealt with in Section 1.2.

1.1.5. Newton's Gravitational Law and the Acceleration of Gravity

Newton proved, mathematically, that the observed motion of the heavenly bodies in the solar system could be accounted for if a force exists between any two masses which is proportional to the magnitude of the masses and inversely proportional to the square of their distances apart. In other words if two bodies, separated by a distance d, have masses m_1 and m_2 there is a force F between them, where

$$F \propto \frac{m_1 m_2}{d^2}$$

The constant of proportionality is Newton's *universal gravitational constant* and is denoted by G. Newton's law may therefore be written

$$F = G\frac{m_1 m_2}{d^2} \qquad \qquad \ldots(1.7)$$

Since a gravitational force exists between any two masses there must be a force between Earth and any body on or near the Earth. This force upon such a body is the quantity which we call its *weight*, and it gives rise to a nearly constant acceleration g if the body is close above Earth's surface and unrestrained.

7

In terms of Newton's second law of motion the gravitational attractive force of Earth upon a body of mass m at, or near, Earth's surface is given by

$$F = mg \qquad \ldots(1.8)$$

and the acceleration of gravity, g, is therefore the attractive force exerted by Earth per unit mass of the body.

If M_E is the mass of Earth and R its radius, and if m is the mass of the body and h its height above Earth's surface, then Equation (1.7) gives

$$F = G \frac{M_E m}{(R+h)^2}$$

and compounding this with Equation (1.8) provides the following expression for g

$$g = G \frac{M_E}{(R+h)^2}$$

The determination of g is a relatively simple matter since it involves the measurement of force per unit mass. Methods include the use of (*a*) inclined planes, (*b*) balls rolling in spherical saucers, (*c*) various types of pendula, (*d*) vertical vibrations of masses on helical springs, and (*e*) free fall of a mass, with electronic timing.

g has different values at different places on Earth's surface. This is due to many causes including lack of homogeneity of Earth itself and the fact that Earth is spheroidal rather than truly spherical, but the main cause is the centrifugal force effect upon a mass of Earth's spin on its own axis, the force being dependent upon latitude.

The determination of G is not easy and involves very delicate measuring techniques.

1.1.6. Elasticity

This relates to the size and/or shape deformation of bodies subjected to external forces and, more particularly, to their recovery of size and/or shape when the deforming forces are removed. The materials of bodies which always entirely recover their size and/or shape on the removal of deforming forces are said to be *perfectly elastic*. Those which exhibit no recovery are termed *perfectly plastic*. Real materials usually exhibit characteristics between these two extremes.

The measures of elasticity are the *moduli of elasticity* which are in turn defined by the ratio of *stress* to *strain*. Stress is simply force per unit area, and it may be normal or tangential to a surface, or it may be in between the two, thereby having components both normal and tangential to the surface. Strains are the deformations produced by stresses and may refer to dimensions or to shape. Dimensional strain may occur only in one dimension, in which case the strain is linear and defined as the ratio $\dfrac{\text{Change in length}}{\text{Original length}}$. On the other hand it may occur in all dimensions in which case it is called bulk strain and is defined as the ratio $\dfrac{\text{Change in volume}}{\text{Original volume}}$. Both of these strains are due to normal stresses.

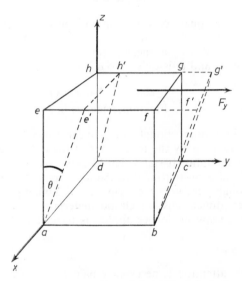

Figure 1.2. Shear strain θ due to shear force F_y

Shape strain, defined by angular, or shear, deformation θ, is illustrated in *Figure 1.2* in which the original volume *abcdefgh* remains unchanged but the shear strain θ involves the new shape *abcde'f'g'h'*. Shear strain is produced by tangential, or shear, forces such as F_y in *Figure 1.2*.

9

The various moduli of elasticity are

1. Young's modulus (E) given by

$$E = \frac{\text{Linear normal stress}}{\text{Linear strain in direction of stress}}$$

2. Bulk modulus (K) given by

$$K = \frac{\text{Compressive or expansive stress}}{\text{Volumetric strain}}$$

3. Rigidity modulus (G) given by

$$G = \frac{\text{Tangential, or shear, stress}}{\text{Shear strain, } \theta}$$

An additional quantity which should be defined is Poisson's ratio. If a specimen of circular cross section has length l and radius r, and it is extended by a linear normal stress it is found that its radius decreases and Poisson's ratio σ is given by

$$\sigma = \frac{\text{Fractional change in } r}{\text{Fractional change in } l}$$

and this ratio is proportional to the linear normal stress.

Hooke's law is really the forerunner of the definitive statement for Young's modulus. It states that 'The extension of a specimen carrying a load is proportional to the load applied'. The constant of proportionality includes the reciprocal of Young's modulus and the specimen dimensions. In elastic materials the limit beyond which Hooke's law no longer applies is termed the *elastic limit*.

1.1.7. Diffusion

Considerable attention is necessarily paid to the phenomenon of diffusion—especially solid state diffusion—in any discussion of Materials Science.

Diffusion is the process whereby two materials in contact will, without outside assistance, interpenetrate due to the effect of their concentration differentials. The process results from the natural energy-reducing tendency of materials to establish a homogeneous distribution of their species throughout a system, which may be recognised as a form of the *second law of thermodynamics*.

If two separate fluids in the same state (liquid or vapour) are brought together they will soon intermix, without the aid of any external mixing agency, until the mixture is homogeneous (assuming the fluids are not immiscible, as are oil and water). The same is also true of solid materials but the process is very much slower. Interpenetration will occur, for example, if two metals are brought into intimate contact.

The principal physical law governing the diffusion of one species through another is Fick's law. Using cartesian coordinate terminology, if $(dn)_x$ is the quantity of a diffusing species which crosses normally the yz plane then Fick's law states

$$(dn)_x \propto A \frac{dc}{dx} dt$$

where A is the area of the plane yz, c is the concentration (quantity per unit volume) of the diffusing species and t is time. Putting a constant of proportionality D into the above gives

$$(dn)_x = -DA \frac{dc}{dx} dt \qquad \ldots(1.9)$$

$$\frac{1}{A} \frac{(dn)_x}{dt} = -D \frac{dc}{dx}$$

$\frac{1}{A} \frac{(dn)_x}{dt}$ can be thought of as the 'flux density' of the diffusing species, in other words the amount of substance per unit time passing normally through unit area. The negative sign simply specifies that the direction of diffusion is towards the lesser concentration of the diffusing species which always travels 'down the concentration gradient'.

The constant of proportionality D is the coefficient of diffusion, or the diffusivity, and a dimensional analysis of the Fick's law equation (1.9) above shows the units of diffusivity to be $m^2 \ s^{-1}$, assuming $(dn)_x$ and dc involve the same units of quantity (e.g. moles).

1.1.8. Viscosity

This phenomenon relates to fluids, that is materials in the liquid or gaseous state. No fluid can resist shear, but all fluids will transmit shear energy in a direction normal to the direction of shear.

Under shear forces all fluids exhibit plastic (completely non-recoverable) deformation.

If a constant shear force in a fluid results in a constant velocity gradient it is described as a *newtonian fluid*. Consider two large parallel plates of area *A*, as in *Figure 1.3*, having a homogeneous fluid between them. Using cartesian coordinates to describe directions in the system let the lower plate be fixed and the upper plate

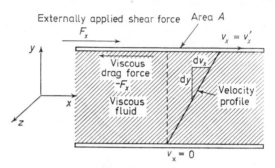

Figure 1.3. *Velocity gradient* dv_x/dy *in viscous fluid due to shear force*

be free and have a constant tangential force F_x applied. When steady state conditions are established the upper plate will move with a constant velocity, v'_x, because the viscous drag force of the fluid on the upper plate will then be $-F_x$. Immediately next to the upper plate the fluid will also move with a constant velocity v'_x whereas immediately next to the lower, fixed, plate the fluid velocity will be zero.

In the fluid there is thus a velocity gradient, described at any point by $\dfrac{dv_x}{dy}$, and in newtonian fluids this is a constant by definition. Newton's law of viscous flow states

$$- F_x \propto A \frac{dv_x}{dy}$$

Putting in a constant of proportionality η gives

$$-F_x = \eta A \frac{dv_x}{dy}$$

and η is termed the coefficient of viscosity (strictly of dynamic

12

viscosity). A dimensional analysis of the above equation shows the units of dynamic viscosity to be kg m^{-1} s^{-1}.

This quantity, η, is based upon unit mass of the viscous medium. Sometimes it is more convenient to work in terms of unit volume in which case the quantity $\mu = \dfrac{\eta}{\varrho}$ (where ϱ is the density of the viscous medium) is used. μ is termed the kinematic (i.e. independent of mass, see Sub-section 1.1.2) coefficient of viscosity. In this case Newton's law is transposed to the form

$$-F_x = \mu A \, \frac{d(\varrho v_x)}{dy}$$

It may be helpful to realise that in the case of dynamic viscosity dv_x represents the increment of momentum per unit mass, whereas in the case of kinematic viscosity $d(\varrho v_x)$ represents the increment in momentum per unit volume.

Of the two equations of viscosity the former, involving η, is probably the more commonly used.

One of the most important practical laws of viscosity is Stokes' law which relates to the terminal velocity v_t of a sphere of density σ and radius r falling through a viscous medium of density ϱ and viscosity η. If F_v be the viscous drag forces on the sphere when it is moving with some velocity v then

$$F_v \propto \eta r v$$

and Stokes found the constant of proportionality to be 6π so that

$$F_v = -6\pi\eta r v \qquad \ldots(1.10)$$

the minus sign merely indicating that F_v is in the opposite direction to v.

When the freely falling sphere reaches its terminal velocity this remains constant at $v = v_t$ by definition and

$$F_{v_t} = -mg \qquad \ldots(1.11)$$

by Newton's second law of motion where m is the mass of the sphere, since no further acceleration takes place. Therefore the relationship

$$mg = 6\pi\eta r v_t \qquad \ldots(1.12)$$

13

is provided. In these expressions m must strictly be the effective mass, in other words the true mass less the mass of displaced viscous medium so that

$$mg = \tfrac{4}{3}\pi r^3(\sigma - \varrho)g \qquad \qquad \ldots(1.13)$$

Then from (1.12) and (1.13) Stokes' law is obtained, which is

$$v_t = \frac{2}{9}\, r^2 \frac{(\sigma - \varrho)}{\eta}\, g$$

1.2. ELECTRICITY AND MAGNETISM

1.2.1. Electrical Quantities and Units

The basic electrical quantity in SI units is electric current (quantity symbol I) whose unit is the ampere (unit symbol A). The unit is defined in Sub-section 1.1.4.

By no means all derived electrical quantities are given here but only those which are considered to be of concern in the following discussions of materials. These quantities and their units are:

(a) *Electrical Quantity or Charge* (symbol Q)—the unit is the coulomb (C) which is the charge associated with a current of one ampere flowing for one second.

$$1\ C = 1\ A\ s$$

(b) *Electrical Potential Difference* (symbol V)—the unit is the volt (V) which is such that if one joule of work is done during the transfer of one coulomb of charge from one place to another then the difference of potential between the two places is one volt.

$$1\ V = 1\ \frac{J}{C}$$

(c) *Electrical Resistance* (symbol R)—the unit is the ohm (Ω) which is such that if, in an electrical conducting medium, an electric current of one ampere flows in response to a potential difference of one volt the resistance of the medium is one ohm.

$$1\ \Omega = 1\ \frac{V}{A}$$

and Ohm's law, $V = IR$, is a useful relationship.

(*d*) *Electrical Field Intensity* (symbol *E*)—the unit is the volt per metre (V m⁻¹) and is that potential gradient which produces a force of one newton on a charge of one coulomb.

$$1\,\frac{V}{m} = 1\,\frac{N}{C}$$

(*e*) *Magnetic Flux Density* (symbol *B*)—the unit is the tesla (T) which is the flux density of that uniform magnetic field which will produce a force of one newton on each metre length of a conductor carrying an electric current of one ampere at right angles to the magnetic field. The force direction is perpendicular to both the field and to the direction of current flow with the sense given by Fleming's left hand mnemonic.

$$1\,T = 1\,\frac{N}{A\,m}$$

Two very useful relationships derive directly from the above electrical and magnetic quantities. From (*d*) it is an easy matter to deduce that the force *F* exerted on an electric charge *Q* which is situated in an electric field of intensity *E* is given by

$$F = EQ \qquad\qquad \dots(1.14)$$

From (*e*) it is easy to deduce that the force *F* on an electrical conductor of length d*s* carrying a current *I* in a magnetic field of flux density *B* is given by

$$F = BI\,ds$$

provided *B* and d*s* are mutually perpendicular. This last relationship can be usefully extended to apply to the force on a particle carrying a charge d*Q* and moving with a velocity *v* in a direction perpendicular to a magnetic field of flux density *B*. From (*a*) above an electric current *I* can be replaced by a rate of charge transfer $\frac{dQ}{dt}$, thus

$$F = B\,\frac{dQ}{dt}\,ds$$

or, alternatively,

$$F = B\,dQ\,\frac{ds}{dt}$$

and since, by definition, $\dfrac{ds}{dt}$ is velocity v the final form becomes

$$F = B\,dQ\,v \qquad \ldots (1.15)$$

Two additional equations complete this sub-section. The first relates the force between two point charges, the magnitude of the charges and their distance of separation. If the force is F and the charges Q_1 and Q_2 are separated by a distance r in free space then

$$F \propto \frac{Q_1 Q_2}{r^2}$$

and this is Coulomb's law. In SI the constant of proportionality is deliberately split into two parts. One part is associated with the spherical symmetry of the geometry of the system and is $\dfrac{1}{4\pi}$, and the other part is associated with the nature of the intervening medium and in this case is the reciprocal of the permittivity of free space (or electric space constant) ε_0. Therefore

$$F = \frac{1}{4\pi\varepsilon_0}\,\frac{Q_1 Q_2}{r^2} \qquad \ldots (1.16)$$

If the intervening medium were to be other than free space the relationship would be

$$F = \frac{1}{4\pi\varepsilon_r\varepsilon_0}\,\frac{Q_1 Q_2}{r^2}$$

where ε_r is the relative permittivity (or dielectric constant) of the intervening medium. The product $\varepsilon_r \times \varepsilon_0$ is termed the absolute permittivity of the medium, symbol ε.

The second of the two equations is that associated with the definition of the unit of electric current, the ampere (*see* Sub-section 1.1.4). This equation relates the force between two parallel electric currents, the length of their conductors and their distance of separation. If the force is F, the two currents I_1 and I_2, the lengths of parallel conductor s and they are separated by a distance r in free space, then

$$F \propto \frac{I_1 I_2}{r}\,s$$

Again the constant of proportionality is deliberately split into two parts. One part, $1/2\,\pi$, is associated with the cylindrical symmetry of the geometry of the system. The other is associated with the nature of the intervening medium and in this case is the magnetic permeability of free space (or magnetic space constant), denoted by μ_0; thus the equation is

$$F = \frac{\mu_0 I_1 I_2 s}{2\pi r} \qquad\qquad \ldots(1.17)$$

If the medium between the two conductors were other than free space the equation would be

$$F = \frac{\mu_r \mu_0 I_1 I_2 s}{2\pi r}$$

where μ_r is the relative magnetic permeability of the medium. The product $\mu_r \times \mu_0$ is termed the absolute magnetic permeability of the medium, symbol μ.

Dimensional analysis of Equations (1.16) and (1.17) shows that the electric space constant ε_0 has dimensions $\dfrac{t^4 I^2}{l^3 m}$ and that the magnetic space constant μ_0 has dimensions $\dfrac{lm}{t^2 I^2}$. The dimensions of the product $\varepsilon_0 \mu_0$ are therefore $\dfrac{t^2}{l^2}$, which are the dimensions of $\dfrac{1}{(\text{Velocity})^2}$, and, in fact

$$(\varepsilon_0 \mu_0)^{-1/2} = c$$

the velocity of electromagnetic radiation in free space.

1.2.2. The Electron Volt and the Atomic Mass Unit

Although not preferred units of SI, electron volts and atomic mass units are so convenient and so widely used in electronic, atomic and nuclear sciences that they are included here and used in the following text. The electron volt is a *unit of energy* for which the symbol is eV, and it is the energy involved in the transfer of one eletronic charge, e (1.602×10^{-19} C), through a potential difference of one volt. Its numerical value in SI units is therefore given by

$$1\ \text{eV} = 1.602 \times 10^{-19}\ \text{J}$$

17

The multiple units keV $= 10^3$ eV, MeV $= 10^6$ eV and GeV $= 10^9$ eV are also in common use.

The atomic mass unit (a.m.u.) is defined as $\frac{1}{12}$th of the mass of an atom of the isotope carbon-12 and its SI equivalent is given by

$$1 \text{ a.m.u.} = 1.660 \ 4 \times 10^{-27} \text{ kg}$$

1.3. WAVE PHENOMENA AND SOME FUNDAMENTALS OF MODERN PHYSICS

1.3.1. Waves

The main parameters of waves, whether they be longitudinal or transverse and whatever their nature, are

(a) wavelength (λ), units m,
(b) period (T), units s, and
(c) amplitude (A), units m

Often, as a matter of convenience, frequency (f)—units hertz (Hz) or s^{-1}—replaces period in discussions of wave phenomena since, by definition,

$$f = \frac{1}{T}$$

and sometimes, also as a matter of convenience, wave number (σ or ν)—units m^{-1}— replaces wavelength, where

$$\sigma \ (\text{or } \nu)| = \frac{1}{\lambda}$$

Waves are periodic phenomena whose displacements (y) are represented graphically with respect to angle (θ) as indicated in *Figure 1.4*, because plane angle is the simplest of all periodic physical quantities. If the wave displacement with respect to distance is under consideration then the distance of one wavelength (λ) is represented by $\theta = 2\pi$ and any other distance x is therefore represented by angle $\theta = \frac{2\pi}{\lambda} x$. If, on the other hand, displacement with respect to time is under consideration then the periodic time T is represented by $\theta = 2\pi$ and any other time t is represented by $\theta = \frac{2\pi}{T} t$.

If the wave form is sinusoidal, the equation

$$y = A \sin 2\pi\left(\frac{t}{T} - \frac{x}{\lambda}\right)$$

represents such a wave travelling in the positive x direction (in cartesian coordinate terminology) and having a velocity given by λ/T. This is a particularly useful equation because any waveform can be represented by an algebraic summation of such sinusoidal (or harmonic) constituents.

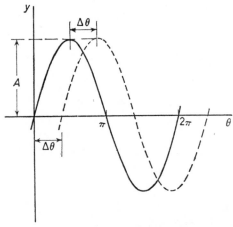

Figure 1.4. Sinusoidal wave forms illustrating phase difference $\Delta\theta$. θ can represent variable distance at fixed time, or variable time at fixed distance

Wave systems of the same physical nature, such as sound waves or electromagnetic waves, combine additively. The maximum resulting effect occurs when the waves are in phase, that is when the systems correspond exactly with respect to time and distance, in such a way that, for example, a crest of one wave occurs at precisely the same time and place as the crests of the others.

If two systems are out of phase by some angle $\Delta\theta$, as between the solid line and pecked line waves in *Figure 1.4*, the resulting effect is less than maximum; and when the two systems are so out of phase that $\Delta\theta = \pi$ $\left(\text{which corresponds to } \Delta x = \frac{\lambda}{2} \text{ and } \Delta t = \frac{T}{2}\right)$

the resulting effect is minimal, and will be zero for waves of the same amplitude.

Representatives of these effects include the phenomena of interference and diffraction which, in the case of electromagnetic waves, are of considerable importance in modern metrology, including the micro-metrology of crystalline structures.

1.3.2. Quantum Ideas

The experimental work of Lummer and Pringsheim on the power emitted by black-body radiators created a dilemma around 1899. The irrefutable experimental relationship they observed between power emitted and radiation wavelength was at variance with that predicted by a rigorous application of existing physical theories. The discrepancy is illustrated in *Figure 1.5*, in which λ is the radiation wavelength and I_λ represents the power emitted per unit area of the black body at wavelength λ.

The dilemma was resolved by Max Planck in 1900 when he introduced the concept of emission of radiant energy in separate

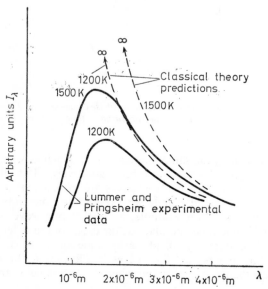

Figure 1.5. Black-body radiation discrepancy between observation and classical theory

units. These units he called *quanta*. In effect Planck suggested that the total power emitted is related to two factors:

(*a*) the energy of each quantum, and

(*b*) the number of quanta emitted per unit time,

such that with decreasing wavelength the individual quanta have increasing energy, generally in accordance with the ideas of the older theories. However, the increasing energy tendency is reversed below a certain wavelength (which depends upon the radiator temperature) by the mathematically more powerful effect of a great decrease in the number of quanta emitted per unit time.

Planck actually made four assumptions in his theory but the two most important and far reaching were

(*i*) radiation is emitted during separate intervals—as 'bursts of energy' called quanta, and

(*ii*) the energy of each quantum, E_q, is proportional to the frequency of its radiation.

From these the first Quantum Physics relationship emerged:

$$E_q \propto f$$

and the constant of proportionality is Planck's constant of 'action' h, having the dimensions of Energy$\times$Time and units J s. Thus the relationship becomes

$$\left. \begin{array}{c} E_q = hf \\[2mm] \text{or} \quad E_q = h\dfrac{c}{\lambda} \end{array} \right\} \qquad \ldots(1.18)$$

where c is the velocity of the radiation and λ its wavelength. The relationship [Equation (1.18)] is called Planck's law.

Soon after this impressive step forward, Einstein in 1905, introduced his special theory of relativity. One of the predictions of this theory is an increase in the observed mass of a body with increasing velocity relative to the observer in accordance with the relationship

$$m = m_0 \left(1 - \frac{v^2}{c^2} \right)^{-1/2}$$

where m is the effective mass of the body for an observer,

m_0 is the mass of the body when it is at rest with respect to the observer,

v is the velocity of the body with respect to the observer, and

c is the velocity of electromagnetic radiation.

Even initially this predicted increase in mass was not without experimental support. Observations of some very fast electrons —produced by certain radioactive materials—had indicated a decrease in their charge-to-mass ratio $\dfrac{e}{m_e}$ compared with slower electrons. However, this was only a qualitatively observed phenomenon.

Later experimenters, and in particular Bucherer, Wolz and Neumann independently between 1908 and 1914, produced conclusive experimental evidence in support of the quantitative relationship

$$m = m_0\left(1 - \frac{v^2}{c^2}\right)^{-1/2}$$

An extremely important equation of modern physics can be derived quite simply from this expression, as follows. Expansion by the binomial theorem gives

$$m = m_0\left\{1 + \frac{1}{2}\frac{v^2}{c^2} + \frac{3}{8}\left(\frac{v^2}{c^2}\right)^2 + \frac{5}{16}\left(\frac{v^2}{c^2}\right)^3 + \ldots\right\}$$

Multiplying throughout by c^2 gives

$$mc^2 = m_0c^2 + \frac{1}{2}m_0v^2\left\{1 + \frac{3}{4}\frac{v^2}{c^2} + \frac{5}{8}\left(\frac{v^2}{c^2}\right)^2 + \ldots\right\}$$

But mc^2 has dimensions $m\dfrac{l^2}{t^2}$ which are the dimensions of energy, therefore one can define *total energy* as E where

$$E = mc^2$$

and this is Einstein's famous mass–energy equivalence relationship.

The total energy is seen to comprise two parts, m_0c^2 which is called the energy of constitution or 'matter energy', and $\dfrac{1}{2}m_0v^2\left\{1 + \dfrac{3}{4}\dfrac{v^2}{c^2} + \ldots\right\}$ the kinetic energy. The results of using this kinetic energy term are very little different from those obtained using the newtonian kinetic energy term $\frac{1}{2}mv^2$ when the velocity of a body is small compared with the velocity of electromagnetic radiation. Newtonian dynamics is by no means rendered obsolete but is merely limited to systems in which velocities are less than a certain value (which depends upon the accuracy desired), for example, $\dfrac{c}{100}$. This is not a great limitation in most everyday engineering.

Combining the two relationships $E_q = hf = h\dfrac{c}{\lambda}$ and $E = mc^2$ suggests some interesting ideas, thus

$$mc^2 = h\frac{c}{\lambda}$$

or
$$mc = \frac{h}{\lambda}$$

Thus a quantum of radiation of wavelength λ has associated with it a quantity $\dfrac{h}{\lambda}$ having the same dimensions as mc which are the dimensions of momentum, and from this point of view a quantum of electromagnetic radiation (a photon) has some of the attributes of a particle.

An extension of the ideas contained in this last equation led the French physicist de Broglie to suggest the application of the equation to any particle of mass m having a velocity v in the modified form

$$mv = \frac{h}{\lambda}$$

thus forecasting a 'wavelike' property for the behaviour of fast moving particles, such that the wavelength is given by $\dfrac{h}{mv}$. The suggestion was eventually shown to be valid by the direct experimental evidence of Davisson and Germer in 1927. This concept of de Broglie's has led to many powerful techniques for the investigation of materials' structure, including those of electron and neutron diffraction, and field-emission and field-ion microscopy.

1.4. QUESTIONS

1. Disregarding the viscous drag forces of air, (*a*) with what velocity must a body be projected vertically upwards to reach a height of 200 m? (*b*) What is the elapsed time from the instant of projection to the instant of reaching 100 m? (*c*) What is the velocity at 100 m?

2. A man-made satellite orbits the Earth in a circular path of radius 7 500 km. What is the periodic time of its revolution? (Take $M_E = 5.98 \times 10^{24}$ kg)

3. The velocity of a transverse wave travelling in a stretched string depends only upon the string length s, the string mass m and the tensional force in the string F. Use the method of dimensions to deduce an expression for the wave velocity v in terms of s, m and F.

4. A small mass of 0.05 kg travels in a circular path of 0.10 m radius with a period 3.14 s. Calculate (a) its speed, (b) its kinetic energy and (c) the centripetal force acting upon the mass.

5. A mass of 2 kg when supported by a wire 1 m long of diameter 10^{-3} m produces an extension 3×10^{-6} m. Calculate Young's modulus for the material of the wire.

6. The supply of oxygen to an immersed electrode in a fuel cell is limited by diffusion across a thin layer of electrolyte next to the electrode surface.

The concentration of oxygen on the outer side of this diffusion layer is 0.30 mol m^{-3}, and is maintained at zero at the electrode surface by the oxygen consuming reaction

$$\tfrac{1}{2} O_2 + H_2O + 2e \rightarrow 2\,OH^-$$

Taking the diffusion layer thickness as 5×10^{-6} m and the coefficient of diffusion of oxygen in the electrolyte as 2.0×10^{-9} m^2 s^{-1}, at what rate is oxygen arriving at the electrode surface?

7. A steel ball-bearing of diameter 2×10^{-3} m falls through a viscous medium of density 1.26×10^3 kg m^{-3} with terminal velocity 6×10^{-3} m s^{-1}. Calculate (a) the coefficient of dynamic viscosity and (b) the coefficient of kinematic viscosity. Take the density of steel to be 7.80×10^3 kg m^{-3}.

8. A very small particle carries an electric charge of 1.60×10^{-19} C and is accelerated from rest through a potential difference of 50 kV. Calculate the kinetic energy of the particle (a) in joules and (b) in electron volts.

9. A particle of mass 9.11×10^{-31} kg carrying an electric charge of 1.60×10^{-19} C is projected with a velocity 2×10^5 m s^{-1} at right angles into a uniform magnetic field of flux density 5×10^{-5} T. What is the path radius of the particle?

10. A plane travelling acoustic wave can be represented by the equation

$$y = 5 \times 10^{-4} \sin (1\,980t - 6x)$$

where y metres is the local displacement of the medium in which the wave travels, t is the time in seconds and x is the distance in metres in the direction of wave propagation. Find (*a*) the vibration amplitude of the medium, (*b*) the vibration frequency, (*c*) the wavelength and (*d*) the velocity of propagation of the acoustic energy in the x direction.

11. What is (*a*) the maximum particle velocity and (*b*) the maximum particle acceleration in a material due to the transmission of an ultrasonic wave of amplitude 10^{-7} m and frequency 50 kHz?

12. What is the energy possessed by a quantum of red light of wavelength 6×10^{-10} m?

13. A quantum of radiation has wavelength 0.1×10^{-10} m. What is the frequency of the radiation?

14. What is the percentage relativistic mass increase of a particle accelerated from rest to a velocity 5×10^7 m s^{-1}?

15. Calculate the length of the de Broglie wave associated with a particle of mass 9.11×10^{-31} kg moving with velocity 10^6 m s^{-1}.

Chapter 2

The Prime Particles

2.1. THE ELECTRON (e)

Electrons can be quite easily obtained for study from, for example, a heated metal filament from which they will be emitted freely.

2.1.1. *The Nature of Electrons*

The precise nature of the electron is not known even today, but for most practical purposes it may be taken to be a negatively charged particle which occupies a spherical space of not more than 10^{-14} m diameter. Although there is some indeterminacy concerning the nature and size of the electron it has certain unvarying physical attributes which may be measured with a high degree of precision. It carries an electric charge, and it exhibits a mass, both of which are extremely small in magnitude.

2.1.2. *The Electronic Charge e*

The electron carries an electric charge which is almost invariably represented by the symbol e. The numerical value of this charge is

$$e = 1.602 \times 10^{-19} \text{ C}$$

and, by convention, the charge is negative.

The really accurate determination of the magnitude of e is due to the American physicist Millikan, the principles of whose method are contained in the following. Using a suitable atomiser, very fine droplets of a non-vaporising oil are introduced into an air space between horizontal parallel metal plates. These plates are connected, through a switch, to a non-fluctuating voltage source to provide a vertical electric field E whose direction and magnitude are under the experimenter's immediate control. The droplets can be observed with an optical arrangement of telescope and illumination.

The mass m of a selected oil droplet is obtained by measuring its terminal velocity when falling freely under gravity in the viscous medium air. For the free fall of a spherical body in a viscous medium Stokes' law states*

$$v_t = \frac{2}{9} r^2 \frac{(\sigma - \varrho)}{\eta} g$$

where v_t = terminal velocity ϱ = density of viscous medium
   r = radius of body η = viscosity of viscous medium
   σ = density of body g = acceleration of gravity

Hence
$$r^2 = \frac{9\eta v_t}{2(\sigma - \varrho)g}$$

If m is the effective mass of the body, that is its real mass less the mass of the displaced viscous medium, it follows that

$$m = \frac{4}{3} \pi r^3 (\sigma - \varrho)$$

$$= \frac{4}{3} \pi (\sigma - \varrho) \left\{ \frac{9\eta v_t}{2(\sigma - \varrho)g} \right\}^{3/2}$$

in which σ, ϱ, η and g are all known quantities, and v_t is observed and measured.

For the determination of e, oil droplets are sprayed into the system and subjected to a short burst of x-rays. X-radiation can cause such droplets to gain or lose electrons so there will then be droplets with various charges, $Q = n \times e$, positive or negative, where n is the number of electrons lost or gained.

The electric field is switched on and varied so as to enable charged droplets to be distinguished from uncharged. A suitably situated charged droplet is selected for continuous observation and is brought to an appropriate 'starting' position using the electric

* Millikan's determination of e showed Stokes' law to be inaccurate for very small spheres—but the inaccuracy is ignored in this discussion.

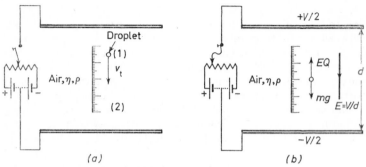

Figure 2.1. The essence of Millikan's method for e.
(a) Free fall of selected droplet (mass m, radius r, density σ) under gravity,
at terminal velocity v_t
(b) Negatively charged droplet stationary: electrostatic force EQ balances
gravitational force mg

field. The field is then switched off so that the free fall terminal velocity v_t can be measured, for example between (1) and (2) in *Figure 2.1 (a)*. This is the observational data necessary for determining the effective mass m of the droplet. By use of the field E the same droplet is returned near to the centre of the experimenter's field of view, and the electric field adjusted so as to hold the droplet absolutely stationary. Under these conditions the electrostatic forces on the droplet $E \times Q$ must be exactly offsetting the gravitational forces on the droplet $m \times g$ and so, as seen in *Figure 2.1.(b)*,

$$mg = -EQ$$

whence

$$Q = ne = -\frac{mg}{E}$$

and

$$e = -\frac{mg}{nE}$$

From a statistically adequate number of observations of Q, which will naturally involve many different values of n, it is taken that the highest common factor is e.

2.1.3. The Electronic Rest Mass m_e

This quantity has never been determined by direct means. Fortunately it is not very difficult to obtain a value for the electronic specific charge, that is the charge per unit mass of the electron, e/m_e. Having obtained the specific charge, and knowing a value for the absolute charge e, it is a simple matter to obtain m_e. From such considerations the value of the electronic mass has been determined as

$$m_e = 9.11 \times 10^{-31} \text{ kg}$$

2.1.4. The Determination of e/m_e

There are many excellent methods for the determination of e/m_e. The essential features of the elegant and accurate one due to Dunnington are presented here.

In *Figure 2.2*, which illustrates the principle of Dunnington's method, $\varkappa$ represents a variable high-frequency oscillator applying alternating potentials across the very closely spaced pairs of electrodes AF and XY. An oxide coated emitter, F, produces thermal electrons in response to indirect heating by the source H. The flux density of a large uniform magnetic field perpendicular to, and into, the plane of the page is represented by B, and the whole apparatus is contained in a highly evacuated enclosure.

During those half-cycles of the high frequency oscillator in

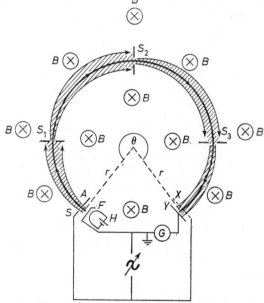

Figure 2.2. Schematic arrangement of Dunnington's method for e/m_e

29

which the electrodes X and A are positive, electrons from F will be accelerated towards A. Some will pass through the aperture S in A and then, being only under the influence of the magnetic field B, will experience a force Bev and describe circular paths of radii $\dfrac{m_e v}{Be}$ $\bigg($ obtained by equating Bev to the newtonian centripetal force $\dfrac{m_e v^2}{r}\bigg)$, v being the velocity of the electrons emerging from A, and having a range of values.

By means of a system of baffle plates, or stops, having very fine and accurately located slits S_1, S_2 and S_3 at a radius r, only those electrons with the specific selected velocity $\dfrac{Ber}{m_e}$ are allowed to complete the circuit from A to X, others being stopped by the baffle plates. By varying B, different velocities can be selected if required.

At the end of their circuit through the complete angle θ the electrons are collected at the electrode Y and are detected by the sensitive galvanometer G, provided they pass the gauze electrode X whose potential with respect to Y is alternating under the influence of $\backsim$.

In determining e/m_e, the frequency (f) of $\backsim$ and/or the flux density of the magnetic field B are adjusted so that there is no deflection of G. Under these conditions no electrons are reaching Y because, after travelling through the complete angle θ the potentials of X and Y are such that the electrons are either directly captured by the electrode X or, if they pass through X and enter the region between Y and X, they are repelled by the local electric field away from Y and back to X where they will be captured.

The above situation of null reading on G demands that the periodic time T of the high-frequency oscillator, or an integral number n times T, be exactly equal to the transit time of the electrons through the angle θ. In other words any electron leaving A when A and X are positive must arrive at X one, two, three or n full cycles later when X is again positive.

Bearing in mind that

$$\text{Velocity} = \frac{\text{Distance}}{\text{Time}}$$

and also that

$$\text{Periodic time } (T) = \frac{1}{\text{Frequency } (f)}$$

it is possible to express the above situation thus:

$$v = \frac{r\theta}{nT} = \frac{r\theta}{n} f \qquad [n \text{ being an integer}]$$

but

$$v = \frac{Ber}{m_e}$$

so

$$\frac{Ber}{m_e} = \frac{r\theta f}{n}$$

whence

$$\frac{e}{m_e} = \frac{\theta f}{Bn}$$

It is quite simple to obtain the value of n; the lowest value of f which gives a null reading of G will be associated with $n = 1$, and successive values of f for null reading of G will be associated with $n = 2, 3, 4$, etc.

The specific charge of the electron, as determined by such means is

$$\frac{e}{m_e} = 1.759 \times 10^{11} \text{ C kg}^{-1}$$

2.2. THE PROTON (p)

Hydrogen is known to consist of protons and electrons, and protons can be obtained for study from hydrogen gas without great difficulty.

2.2.1. The Nature of Protons

The proton is generally accepted as being a particle. Although it is a massive particle compared with the electron it is, nevertheless, extremely small, having an effective diameter of around 6×10^{-15} m. Like the electron, the proton has certain properties which are in no doubt. It carries an electric charge and it exhibits a mass.

2.2.2. The Proton Charge

It is known that the smallest unit of hydrogen comprises one proton and one electron; it is also known that this unit carries no charge. From this information it is inferred that the proton carries a charge exactly equal in magnitude to that carried by the electron, so its magnitude can also be represented by e, but of course it is of opposite sign.

31

2.2.3. The Proton Mass m_p

As in the case of the electronic mass the proton mass, m_p, cannot be obtained directly. It has to be obtained indirectly by measuring e/m_p, the specific proton charge, and then deduced using the inferred value of the proton absolute charge e.

The accepted value for the proton mass is

$$m_p = 1.673 \times 10^{-27} \text{ kg}$$

Thus m_p is about 1 836 times the magnitude of m_e.

2.2.4. The Determination of e/m_p

There are several methods for obtaining the specific proton charge e/m_p. The parabola method due to J. J. Thomson will be considered here because of the important principles involved, but a later apparatus is described in Sub-section 3.2.2.

In Thomson's method (*see Figure 2.3*), protons are produced from gaseous hydrogen in a low pressure discharge tube. Between

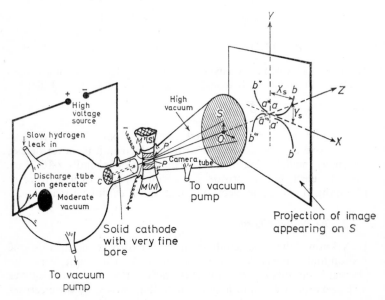

Figure 2.3. General arrangement of J. J. Thomson's apparatus for positive ion mass determination

the electrodes A and C there is a high voltage resulting in a strong electric field, and a stream of protons is thus accelerated towards cathode C, which consists of a metal bar with a very fine uniform central bore down which passes a high velocity stream of these protons. This emerges as a fine, collimated, proton beam into the second stage of the apparatus referred to as the *camera*.

In the discharge tube it is sufficient to maintain only a moderately low pressure of hydrogen gas, say 10 N m^{-2}, whereas in the camera it is necessary to have a much lower pressure, say 10^{-2} N m^{-2}. This pressure differential between the two parts of the apparatus necessitates a rather elaborate arrangement of vacuum pumps, a carefully controlled hydrogen inlet leak and a long, very fine, bore in the cathode.

Uniform parallel magnetic and electric deflecting fields having the same dimensions are produced in the camera by the electromagnet MM', whose pole pieces P and P' are electrically insulated from the remainder of the electromagnet core and can therefore also act as the electrodes for the production of the electric field. At the end of the camera tube is a photosensitive screen S which becomes exposed where it is subject to bombardment by the protons.

When neither the magnetic nor the electric field is switched on in the camera, the stream of protons is undeviated and a central spot O appears on the photosensitive screen. If, alone, an electric field of strength E is switched on in the sense indicated by the electrical polarity of P and P' in *Figure 2.3* (that is in the positive Y direction of the illustrated left-handed system of coordinates), the proton beam will be deflected in the positive Y direction. Conversely, if only a magnetic field of flux density B is switched on in the sense shown by the magnetic polarity of M and M' in *Figure 2.3* (again in the positive Y direction) the proton beam will be deflected horizontally in the positive X direction (looking at the screen from inside the camera).

When both fields are applied simultaneously the trace that appears on the screen S is the parabolic arc represented by the projection ab.

'Mirror images' of this arc, such as $a'b'$, $a''b''$ and $a'''b'''$, can be produced by appropriate reversals of the electric and/or magnetic deflecting fields. Measurement of any pair of coordinates of the parabolic arc, together with a knowledge of the camera tube dimensions and the electric and magnetic deflecting fields, enables e/m_p to be evaluated, as will be understood from the following.

For any proton entering the deflecting fields with some velocity

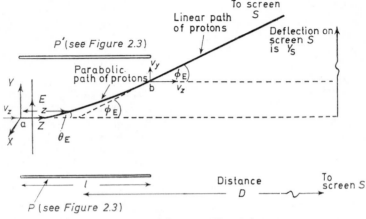

Figure 2.4. Electric deflection in Thomson's apparatus

v_z, the electrical and magnetic deflections can be treated separately. Consider first the effect of the electrical field only. The proton enters the field at a, see *Figure 2.4*, with only a forward velocity v_z and it emerges from the field at b with its forward velocity unchanged (in accordance with Newton's first law of motion), but it has gained an additional component of velocity v_y imparted by the deflecting field E.

Since the forward velocity v_z is unchanged, if the forward distance travelled in the field is z, the time spent travelling this distance is $\dfrac{z}{v_z}$. One of the kinematic space–time relationships is

$$s = ut + \tfrac{1}{2}at^2$$

and this can be modified to suit the motion in the Y direction of the case under consideration, as follows

$$y = u_y t + \tfrac{1}{2}a_y t^2$$

Initially there is no component of velocity in the Y direction so $u_y = 0$ (proton at a in *Figure 2.4*) and the relationship becomes

$$y = \frac{1}{2} a_y t^2$$

$$= \frac{1}{2} a_y \left(\frac{z}{v_z}\right)^2 \qquad \ldots(2.1)$$

34

Applying Newton's second law of motion in the form 'Force = Mass$\times$Acceleration' gives

$$Ee = m_p a_y$$

whence

$$a_y = \frac{Ee}{m_p}$$

and this in conjunction with the kinematic Equation (2.1) above gives

$$y = \frac{1}{2} \frac{Ee}{m_p} \left(\frac{z}{v_z}\right)^2$$

At any distance z in the field the angle θ_E is given by $\dfrac{dy}{dz}$

thus

$$\theta_E = \frac{dy}{dz}$$

$$= \frac{Ee}{m_p} \frac{z}{v_z^2}$$

The angle ϕ_E is associated with $z = l$

thus

$$\phi_E = \frac{Ee}{m_p v_z^2} l$$

In a well designed apparatus the deflection Y_S on the screen S is given, to a good approximation, by $D\phi_E$

so

$$Y_S \approx \frac{DEel}{m_p} \frac{1}{v_z^2} \qquad \ldots (2.2)$$

Now consider magnetic deflection only, on the same proton, and represented in *Figure 2.5*. Since the magnetic field B is vertically upwards the deflection produced on the screen S will be in the X direction. The analysis of this deflection is as follows.

Newtonian dynamics show that a force of constant magnitude, always acting in the same plane and at right angles to the instantaneous velocity vector of a body travelling with constant speed, will cause that body to travel in a circular path. Such a force is exerted on the proton by the magnetic field B. The force has magnitude Bev_z where e is the proton charge, and v_z is its speed. The laws governing circular motion require a body of mass m_p

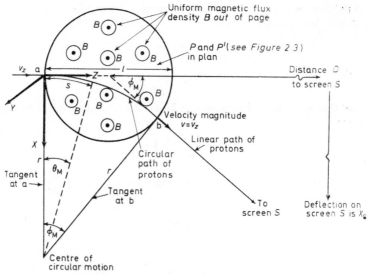

Figure 2.5. Magnetic deflection in Thomson's apparatus

travelling with speed v_z in a circular path of radius r to have a centripetal force of magnitude $\dfrac{m_p v_z^2}{r}$ acting upon it. Thus

$$Bev_z = \frac{m_p v_z^2}{r}$$

from which

$$r = \frac{m_p v_z}{Be}$$

At any arc length s from a into the field the angle θ_M is given by

$$\theta_M = \frac{s}{r}$$

$$= \frac{Be}{m_p}\,\frac{s}{v_z}$$

The angle ϕ_M is associated with $s =$ arc ab, but the arc ab is very nearly equal to l

thus

$$\phi_M \approx \frac{Be}{m_p}\,\frac{l}{v_z}$$

36

In a well designed apparatus the magnetic deflection X_S on the screen S will be very nearly $D\phi_M$

so
$$X_S \approx \frac{DBel}{m_p} \frac{1}{v_z} \qquad \ldots (2.3)$$

By eliminating the variable v_z from Equations (2.2) and (2.3) it is a simple matter to obtain the relationship

$$Y_S = \frac{m_p E}{DeB^2 l} X_S^2$$

which is the equation of the parabola obtained on the screen S. Rearrangement of this gives

$$\frac{e}{m_p} = \frac{E}{DB^2 l} \frac{X_S^2}{Y_S}$$

2.3. THE NEUTRON (n)

Neutrons are not so easily obtained, or detected, as electrons and protons and consequently were not discovered until some time after the other two particles. Neutrons have to be obtained from quite complex nuclear reactions.

2.3.1. The Nature of Neutrons

The neutron is a massive particle comparable in size to the proton (diameter around 10^{-14} to 10^{-15} m) and of a similar mass. It carries no electric charge, and this in itself makes its detection difficult.

2.3.2. The Neutron Mass m_n

As for electrons and protons, the direct determination of neutron mass has not yet proved possible. Since the neutron itself carries no charge, methods directly involving specific charge cannot be used. Indirectly, however, they can: the charge to mass ratio is determined for a known, charged, combination of protons, neutrons and electrons and from this the neutron mass is deduced. The accurate evaluation of m_n by such means is not straightforward and involves considerations which are discussed in Section 3.4.

The accepted mass of the neutron is

$$m_n = 1.675 \times 10^{-27} \text{ kg}$$

which is slightly greater than that of the proton.

2.4. OTHER PARTICLES

The electron, proton and neutron are the commonplace atomic particles but there are many other particles whose existence is generally only of concern in the study of high energy physics, and which are therefore of comparatively little interest in an introductory study of materials. For the sake of completeness, however, Table 2.1 lists some of the better known of these other particles, together with their mass, charge and symbol.

Table 2.1

Name	Mass relative to electron rest mass	Charge	Symbol
Neutrino	$< \dfrac{1}{1\,000} \times m_e$	0	ν
Positron	$1 \times m_e$	$+e$	β^+
Muon or μ-meson	$207 \times m_e$	$+e$ or $-e$	μ^+ or μ^-
Pion or π-meson	$273 \times m_e$	$+e$ or $-e$	π^+ or π^-
K-meson	$966 \times m_e$	$+e$ or $-e$	K^+ or K^-
Various Hyperons	about 2 100 to $2\,600 \times m_e$	$+e$, 0 or $-e$	Various

2.5. THE PHOTON (γ)

Although not a particle in the classical sense of being an entity with a definite size, shape and mass, the photon is nevertheless sometimes conveniently treated as a particle.

The photon is the name given to the unit, or quantum, of electromagnetic radiation and its particulate nature was first suggested by Einstein. In Chapter 1 it was seen that a quantum of electromagnetic radiation having wavelength λ has an associated quantity $\dfrac{h}{\lambda}$ with the same dimensions as momentum. It is known experimentally that a stream of photons (such as a beam of light) will exert a pressure when it falls on matter. It is also known experimentally that photon–particle collisions can occur in which energy and momentum are conserved in the manner applicable to elastic collisions between conventional particles.

Thus from many viewpoints it is convenient to think of the photon as a particle of zero mass and zero charge yet possessing energy hf and momentum $\dfrac{h}{\lambda}$.

2.6. QUESTIONS

1. In a Millikan apparatus for determining the electronic charge e, the plates are 10^{-2} m apart. An oil droplet of mass 1.63×10^{-15} kg is maintained stationary by a potential difference of 500 V across the plates. Calculate a value for the charge carried by the oil droplet disregarding flotation forces. How many electrons does this represent?

2. In a Millikan apparatus a charged oil droplet fell under gravity alone with an observed terminal velocity of 4×10^{-5} m s^{-1}. Then, under the influence of a field of 2×10^5 V m^{-1} it rose with an observed terminal velocity 1.2×10^{-6} m s^{-1}. Given that the radius of the droplet was 2.3×10^{-6} m and the dynamic viscosity of the air was 1.80×10^{-5} kg m^{-1} s^{-1}, calculate a value for the charge on the droplet, disregarding flotation forces.

3. In a Dunnington apparatus for determining $\dfrac{e}{m_e}$, the angle subtended by the two pairs of electrodes at the centre, θ, is 270°. The magnetic flux density is adjusted to 5×10^{-5} T. The galvanometer indicates zero current for applied frequencies of 5.60 MHz, 3.73 MHz and 1.87 MHz, the last of these being the lowest value to give a null reading on the galvanometer. Calculate a value for $\dfrac{e}{m_e}$.

4. An electron beam is produced in a vacuum by accelerating electrons from rest through a potential difference of 1.5 kV. This beam is deflected between parallel plates 30 mm long and 7.5 mm apart with a potential difference of 150 V between them. What is the angular deflection of the beam?

5. A cathode ray oscilloscope provides a beam of electrons of velocity 2.7×10^7 m s^{-1}. Find the flux density of a magnetic cross field, extending over 50 mm of the beam path and whose centre is 0.40 m from the screen, which will give a linear deflection of 10 mm at the screen.

6. Compare the de Broglie wavelength associated with electrons accelerated from rest through a potential difference of 1 000 V with that associated with protons accelerated from rest through the same potential difference.

7. In a Thomson apparatus, such as that shown in *Figure 2.3*, the electric and magnetic deflecting fields are respectively 2×10^3 V m^{-1} and 10^{-2} T. The screen is 0.30 m from the centre of the deflecting system, and the deflecting fields are 2×10^{-2} m long.

Using the apparatus to determine the specific charge of protons and hence the proton mass, the coordinates of a point on a parabola were found to be $X_S = 30$ mm and $Y_S = 31.3$ mm. (*a*) Calculate a value for the specific charge carried by the proton, then (*b*) using the known proton charge, deduce the proton mass.

8. For the system described in Question 7, what is (*a*) the velocity of the protons whose coordinates on the screen are given, and (*b*) the de Broglie wavelength of these protons?

Chapter 3

Elementary Assemblies of the Prime Particles

3.1. THE FREE ATOM

All matter can be considered simply as different assemblies and arrangements of the three prime particles, electron, proton and neutron. The various units of assembly are discussed here.

3.1.1. The Elements

The most elementary assemblies of prime particles are electrically neutral and are called *atoms*. There are many different atoms and their most convenient main classification involves the number of protons they contain. In this classification there are only 92 different natural atoms, although others can be produced artificially. These 92 are referred to as the natural *elements*.

The simplest atom contains one proton and is the atom of the element called hydrogen (chemical symbol H), the next simplest contains two protons and is the atom of the element helium (chemical symbol He) and so on, with increasing numbers of protons, to the most complex natural element uranium (chemical symbol U), whose atom contains 92 protons.

41

3.1.2. Atomic Number (Z)

As pointed out in Chapter 2, the proton carries a positive charge equal in magnitude, but opposite in sign, to the electronic charge. Normally, atoms have no charge because they contain as many electrons as they do protons, and the electron and proton charges balance each other.

Because the number of protons in an atom classifies the element —as also does the number of electrons in the normal atom—this quantity is called the *atomic number*. Thus in addition to its chemical name (such as hydrogen, helium, lithium or uranium) an element is identified by its atomic number, for example 1 (hydrogen), 2 (helium), 3 (lithium), and so on to 92 (uranium).

3.1.3. Mass Number (A) and Isotopes

So far no mention has been made of neutrons in the atom. Atoms may, in fact, contain any of several different numbers of neutrons without in any way altering their identity as a particular element. For example some hydrogen atoms have no neutrons, some have one neutron and it is possible to produce others which have two neutrons, but in all cases they remain hydrogen by definition, that is the element of atomic number 1.

Reference to the numerical values for the masses of electrons, protons and neutrons shows that the proton and neutron are both over 1 800 times more massive than the electron, so that practically the entire mass of any atom is contained in its protons and neutrons. The number of protons and neutrons in an atom defines its *mass number*.

Hydrogen, atomic number 1, can therefore be said to exist in three atomic forms defined by mass number 1 (no neutrons), mass number 2 (1 neutron) and mass number three (2 neutrons). In a similar way uranium, atomic number 92, can exist in several forms; for example one form has mass number 236 (144 neutrons) and another form has mass number 238 (146 neutrons).

Such different mass forms of the same element are referred to as *isotopes* of that element. An element is completely specified by its atomic number, but an isotope is completely specified only when both atomic number and mass number are known.

Elementary Assemblies of the Prime Particles

3.1.4. Symbolic Notation

The generally accepted symbol for atomic number is Z and that for mass number is A, and $A - Z$ therefore gives the number of neutrons in an atom of the isotope concerned.

A useful notation in atomic and nuclear physics, and one which completely specifies an isotope, is

$$^A_Z\text{Chemical formula}$$

For example, ^1_1H identifies the hydrogen isotope having no neutrons, ^2_1H the hydrogen isotope with one neutron (often called deuterium), and ^3_1H that with two neutrons (often called tritium). In the same way $^{236}_{92}\text{U}$ identifies the uranium isotope with 144 neutrons and $^{238}_{92}\text{U}$ that with 146 neutrons.

The chemical formula is really superfluous in this notation since it provides exactly the same information as does the atomic number Z, namely the identity of the element. In atomic and nuclear physics, the chemical symbol is therefore sometimes omitted and the above examples could then be written $^1_1$, $^2_1$, $^3_1$, $^{236}_{92}$, and $^{238}_{92}$; or perhaps in parentheses thus $\binom{1}{1}$, $\binom{2}{1}$, etc. On the other hand, the symbol Fe, for example, is much more commonly recognised as representing iron than is the plain number 26, and so the chemical formula still tends to be used in the symbolic notation except in specialist texts.

3.1.5. Ions

Some atoms can, without great difficulty, be made to gain, and others to lose one, two, three or even four electrons (sometimes more, but this requires increasing amounts of energy). This process has only a small effect on the mass of the atom but it does convert it from an uncharged to a charged particle, positively charged if it has lost electrons and negatively charged if it has gained them.

Such charged particles are no longer referred to as atoms but as *ions*, and the process of their formation is called *ionisation*. For example, atoms of the isotope $^{56}_{26}\text{Fe}$ are easily made to lose two electrons to become iron ions, each with a charge $+2e$; conversely, atoms of the isotope $^{16}_8\text{O}$ can be made to accept two additional electrons each and so become oxygen ions with charge $-2e$. Elements do not form both types of ion with equal facility but are more or less restricted to forming either positive or negative ions, although there are a few which readily form positive ions in some circumstances and just as readily form negative ions in others.

Positive ions are formed by the metallic elements, and in fact the ready formation of positive ions is an important factor in classifying metals, examples of metals being sodium, magnesium, iron, copper, zinc, silver, etc. Negative ions are formed by the non-metals such as nitrogen, oxygen, fluorine, phosphorus, sulphur and chlorine, etc. The 'in between' elements which form either positive or negative ions include carbon, silicon and germanium and these are all very important materials in modern technology.

To carry the symbolic notation given in Sub-section 3.1.5 one stage further, it is possible to identify unambiguously a specific ion of a particular isotope of an element by indicating its charge as well as both its atomic and mass numbers. This is done by adding a post-superscript to the isotope symbol. The doubly charged ion of the isotope of iron which contains 30 neutrons can thus be identified by any of the symbols

$$^{56}_{26}\text{Fe}^{++}, \left(^{56}_{26}\right)^{++}, \text{ or } \left(^{56}_{26}\right)^{2+}$$

and the triply charged ion of the same isotope by

$$^{56}_{26}\text{Fe}^{+++}, \left(^{56}_{26}\right)^{+++} \text{ or } \left(^{56}_{26}\right)^{3+}$$

To give an example of a negative ion, that of the oxygen isotope which is doubly charged and which contains 8 neutrons can be identified by

$$^{16}_{8}\text{O}^{--}, \left(^{16}_{8}\right)^{--}, \text{ or } \left(^{16}_{8}\right)^{2-}.$$

Positive ions are frequently called *cations* and negative ions *anions* because in electrolysis, and in gaseous discharge tubes, cations move towards the cathode and anions towards the anode.

3.2. ATOMIC MASS DETERMINATION

3.2.1. General

The first effective apparatus for atomic mass determination (mass spectrometer) was that of J. J. Thomson, which has already been dealt with in relation to the determination of the proton mass (*see* Sub-section 2.2.4). For the determination of other masses using Thomson's apparatus, the discharge tube in which ionisation occurs is made to contain a low pressure vapour of the material concerned, and pairs of coordinates of the parabolas so obtained are measured in the manner already described.

The Thomson apparatus is difficult to use, and its ability to distinguish masses which are nearly the same is not great, nevertheless it was this instrument which first led to the discovery of isotopes by resolving two different mass forms of natural neon, $^{20}_{10}$Ne and $^{22}_{10}$Ne.

A later mass spectrometer, and one having high powers of resolution, was produced by F. W. Aston who used it to carry out a large number of mass determinations, thereby adding greatly to the scientific knowledge of his day. A yet more recent mass spectrometer is that due to Bainbridge which is the one dealt with in some detail here. Various forms of this instrument are in wide use in industry today.

3.2.2. The Bainbridge Mass Spectrograph

In this instrument a stream of ions of known charge Q (which may be $\pm e$, $\pm 2e$, or $\pm 3e$, for example, but is of the same sign on each ion) whose masses are required is filtered to provide a beam consisting only of ions with a particular velocity. This filtered ion beam is projected at right angles into a uniform magnetic field. If M is the mass of any such ion, v the velocity transmitted by the filter and B the flux density of the uniform magnetic field, then in the magnetic field the electromagnetic force acting on the ion is BQv, producing circular motion of radius R such that

$$\frac{Mv^2}{R} = BQv$$

This can be transposed to give

$$M = \frac{BQR}{v} \qquad \ldots(3.1)$$

The essential features of the instrument are shown in *Figure 3.1*. In the velocity filter, or selector, there are crossed electric and magnetic fields (E_s and B_s respectively) which produce opposing deflecting forces on any charged particles passing through them. Only those particles will emerge from the slit for which the electric deflecting force is exactly balanced by the magnetic deflecting force, that is for which

$$E_sQ = B_sQv$$

45

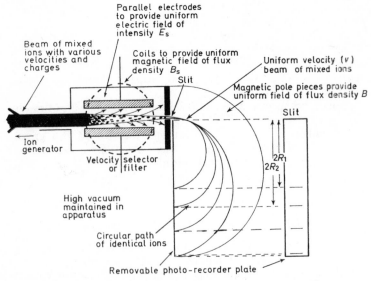

Figure 3.1. Main features of Bainbridge's mass spectrograph

This reduces to

$$v = \frac{E_s}{B_s} \qquad \qquad \ldots(3.2)$$

which is single valued for constant E_s and B_s, and independent of M and Q.

Values of R are obtained from the photo-recording plate of the spectrograph. Hence M is determined by using Equations (3.1) and (3.2) which combine to give

$$M = BQR \frac{B_s}{E_s}$$

In this expression the only quantity on the R.H.S. whose value will be in any doubt is Q. However Q must be $\pm ne$, where n is an integer and probably 1, 2 or 3. This apparent uncertainty is easy to resolve in practice.

46

3.2.3. Atomic Mass Units

These units are commonly used in atomic and nuclear physics and in materials science rather than straight SI units. The SI equivalent of the atomic mass unit is given by

$$1 \text{ a.m.u.} = 1.660\ 4 \times 10^{-27} \text{ kg}$$

The a.m.u. is therefore not greatly different from the mass of the proton or the neutron.

Prior to 1961, the atomic mass unit was defined to be $\frac{1}{16}$th of the mass of the atom of the oxygen isotope $^{16}_{8}O$ (the most abundant in natural oxygen). In 1961 however, the International Union of Pure and Applied Physics (IUPAP) and the International Union of Pure and Applied Chemistry (IUPAC) agreed to adopt a unified scale of atomic masses based upon the mass of the atom of the carbon isotope $^{12}_{6}C$, and the atomic mass unit is now defined as $\frac{1}{12}$th of this.

Numerical differences in atomic masses as defined on the old and the new scales are only about 40 parts per million and therefore mainly of academic rather than practical significance.

3.2.4. Atomic Weight

This term, a hangover from early methods of determination, strictly relates to mass and is inappropriate in modern science. A more suitable term is that recently recommended in relation to the International System of Units (SI) and is *Relative Atomic Mass*.

In nature all elements exist as mixtures of the atoms of several of their various possible isotopes—although in some cases there is a preponderance of one particular isotope. The relative atomic mass (atomic weight) of an element is that number which expresses in atomic mass units the average mass per atom of the naturally occurring substance.

As an example, although neon (Ne) has several possible isotopes, in natural neon the only three of any significant abundance are $^{20}_{10}Ne$ (90.92% abundance), $^{21}_{10}Ne$ (0.26% abundance) and $^{22}_{10}Ne$ (8.82% abundance). If these are found to have masses 19.996 amu, 20.994 amu and 21.992 amu respectively by mass spectrometry, then taking 100 parts by weight of natural neon the average mass is given by

$$\frac{(90.92 \times 19.996) + (0.26 \times 20.994) + 8.82 \times 21.992)}{100} = 20.175 \text{ amu}$$

The often quoted value of 20.183 is the 1956 international value and was based upon the early chemical unit of atomic weight—$\frac{1}{16}$ th of the average mass per atom of naturally occurring oxygen (comprising $^{16}_8O$, $^{17}_8O$ and $^{18}_8O$). The value currently recommended is that published by IUPAC in 1967. and is 20.179 ± 0.003; this is, of course, relative to the isotope $^{12}_6C$ and based upon the best available measurements at that date.

3.3. THE PHYSICAL MODEL OF THE ATOM

3.3.1. General

Probably the most useful physical model of the atom is that due to Lord Rutherford and Niels Bohr. Based upon experimental evidence Rutherford proposed that the atom comprises a small, dense, positively charged central nucleus, surrounded by negative charge contained in a spherical volume which is mostly empty space. This concept was extended such that the negative charge is considered as a planetary system of negative particles (electrons) orbiting about the central nucleus.

By making several assumptions, of which two were quite arbitrary, Bohr was able to produce a mathematical argument in support of the above planetary model in the case of the lightest, and therefore the simplest atom, hydrogen. The main achievement of Bohr's theory lay in providing a very elegant explanation for the principal features of the emission spectrum produced by hydrogen in a low pressure gas discharge tube.

3.3.2. The Hydrogen Spectrum

It had been known for many years (since 1884) that the emission spectrum of hydrogen—the least complex of all elemental spectra—comprised lines representing definite wavelengths (λ), which fitted the expression

$$\frac{1}{\lambda} = R\left(\frac{1}{a^2} - \frac{1}{b^2}\right)$$

the 'Balmer–Rydberg–Ritz–Paschen' formula in which R is an experimental quantity, the Rydberg constant ($1.097\,373 \times 10^7$ m^{-1}).

If a is made 2 and b successively 3, 4, 5,..., in this formula a series of values for λ is obtained which exactly describes the first observed hydrogen spectral lines, the Balmer series (which extends

48

across the visible waveband into the ultra-violet). This series is illustrated in *Figure 3.2* in which the principal lines are given their usual symbols H_α, H_β, H_γ and H_δ.

Making a equal 1 and b successively 2, 3, 4, ..., describes another series (the Lyman series) of spectral lines, but not in the visible waveband. Again, making a equal 3 and b successively 4, 5, 6, ..., provides the Paschen series; making a equal 4 and b successively 5, 6, 7, ..., provides the Brackett series and making a equal 5 and b successively 6, 7, 8, ..., provides the Pfund series.

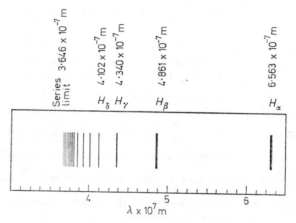

Figure 3.2. The Balmer spectral lines of hydrogen

3.3.3. The Bohr Theory for Hydrogen

As indicated above, several assumptions are necessary to Bohr's theory of the hydrogen atom and these are essentially as follows.

(1) Coulomb's law for the force between charged bodies applies on the atomic scale of dimensions thus:

$$F \propto \frac{Q_1 Q_2}{r^2}$$

where F is the force, Q_1 and Q_2 are the charges on the bodies and r is the separation of the centres of charge of the bodies. In SI units the constant of proportionality has been shown to be $\dfrac{1}{4\pi\varepsilon_0}$ where

ε_0 is the permittivity of free space (or electric space constant), so

$$F = \frac{1}{4\pi\varepsilon_0}\frac{Q_1 Q_2}{r^2}$$

and the sense of F is, of course, attraction if Q_1 and Q_2 are of opposite sign and repulsion if they are of the same sign.

(2) Newton's laws of motion are applicable on the atomic scale of dimensions. For example, a body of mass m_e moving in a circular path of radius r with speed v must have a force of constant magnitude $\dfrac{m_e v^2}{r}$ acting upon it always towards the centre of the circle.

(3) The single electron of the hydrogen atom revolves in circular orbits about a single proton nucleus, and the centre of mass of the system is assumed to be at the centre of the proton (p).

(4) The *only* orbits available to the electron are those for which the angular momentum (i.e. the moment of momentum) $m_e vr$ has the specific values $n\times\dfrac{h}{2\pi}$, where n is an integer (1, 2, 3, 4, ..., etc.) and is called the *principal quantum number*, and h is, of course, Planck's constant of action. This condition $m_e vr = \dfrac{nh}{2\pi}$ is often referred to as the 'quantisation of the angular momentum' and is quite arbitrary in Bohr's treatment.

(5) The electron can make transitions from one available orbit to another (i.e. it can make 'quantum jumps') and the quantity of energy associated with such transitions must be either supplied to, or emitted from the atom. Energy can be supplied to the atom in any form such as heat, light or mechanical energy, but it can only be emitted from the atom as electromagnetic radiation, in other words as photons, whose energy to wavelength relationship derives from Planck's law ($E = hf$) and is

$$E = \frac{hc}{\lambda}$$

Of the above assumptions, (4) and (5) were really revolutionary and are the very essence of the Bohr theory. However, a little thought reveals that assumptions (1) and (2) are also very remarkable in that laws which govern behaviour in one case for astronomical dimensions and in the other case for large scale laboratory and engineering dimensions are assumed to hold good in an atomic system in which the dimensions are anything from ten to twenty orders of magnitude less.

Assumption (3) is merely a contribution to mathematical simplicity.

Consider the electron of charge e and mass m_e revolving in a circular orbit of radius r round a single proton nucleus whose charge is represented by $+e$. In line with assumption (3), the nucleus is taken to be so massive compared with the electron that the centre of mass of the system is assumed to be at the centre of the nucleus. *Figure 3.3* illustrates the model under consideration.

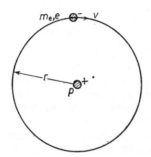

Figure 3.3. Model of H atom for Bohr theory

The centripetal force for this circular motion must be provided by the electrostatic Coulomb attraction between the electron and the proton, and this can be written

$$-\frac{m_e v^2}{r} = \frac{-e \cdot +e}{4\pi\varepsilon_0 r^2}$$

the sign convention for the newtonian force being that a positive force is in the direction of increasing r. The above relationship can be rearranged as

$$m_e v^2 r = \frac{e^2}{4\pi\varepsilon_0} \qquad \ldots(3.3)$$

Bohr's first arbitrary assumption, number (4) above, gives

$$m_e v r = n\frac{h}{2\pi} \qquad \ldots(3.4)$$

Squaring Equation (3.4) and dividing the result by Equation (3.3) the following is obtained:

$$r = n^2 \frac{h^2 \varepsilon_0}{\pi m_e e^2}$$

In other words there is a series of possible values for r which increases as the square of the natural numbers, and the innermost orbit has radius r_1 associated with $n = 1$, thus

$$r_1 = \frac{h^2 \varepsilon_0}{\pi m_e e^2}$$

and

$$r_2 = (2)^2 \times r_1$$

$$r_3 = (3)^2 \times r_1$$

$$r_4 = (4)^2 \times r_1$$

etc.

Since h, π, ε_0, m_e and e are all known physical or mathematical constants, the above expression for r_1 enables a good estimate to be made of the minimum, normal, size of the hydrogen atom, which is 5.292×10^{-11} m radius.

In order to implement the proposals in Bohr's second arbitrary assumption, number (5) above, it is necessary to determine the energy differences between the various orbits available to the electron. Both kinetic energy (E_k) and potential energy (E_p) must be taken into account because the total energy includes both of these, but matter energy (see Sub-section 1.3.2.) can be omitted since it is invariant for the electron in any orbit. The kinetic energy is easily obtained, and is

$$E_k = \frac{1}{2} m_e v^2 \qquad \ldots \text{(from Newton)}$$

$$= \frac{+e^2}{8\pi \varepsilon_0 r} \qquad \ldots \text{[from Equation (3.3)]}$$

The potential energy of a body is the work done *on* the body in moving it from a position where the potential energy is *defined* to be zero (arbitrarily) to that position at which the potential energy is required.[!] By definition, a free electron (that is one which is infinitely far from any charged particle or body) has zero potential energy.

The magnitude of the work done in moving the electron from infinity to a distance r from the nucleus is

$$|E_p| = \left| \int_\infty^r F \, dr \right|$$

where the mathematical symbol | | refers to the 'modulus' of the contained quantity, that is its magnitude only without any reference to its sign.

Coulomb's law gives

$$F = \frac{-e^2}{4\pi\varepsilon_0 r^2}$$

so

$$|E_p| = \left| \int_\infty^r \frac{e^2}{4\pi\varepsilon_0 r^2} \, dr \right|$$

$$= \left| \frac{e^2}{4\pi\varepsilon_0} \left[\frac{1}{r} \right]_\infty^r \right|$$

$$= \left| \frac{e^2}{4\pi\varepsilon_0 r} \right|$$

A little thought should make it clear that, because the force between electron and nucleus is attractive, the electron has not had work done *on* it in moving from ∞ to r but that work has been done *by* the electron. In other words the work done *on* the electron in moving from ∞ to r is negative. The value of the potential energy for the electron in an orbit of radius r is thus given by

$$E_p = -\frac{e^2}{4\pi\varepsilon_0 r}$$

The fact that the potential energy is negative has no fundamental significance, it simply depends upon the arbitrary choice of zero for E_p.

Expressions have now been obtained for both kinetic energy (E_k) and potential energy (E_p), and the total energy E is the sum of these two thus

$$E = E_k + E_p$$

$$= -\frac{e^2}{8\pi\varepsilon_0 r}$$

5*

53

Since an expression has already been obtained for r which contains only known physical constants, this can be put into the above to give a final expression for the total energy E, which is

$$E = -\frac{1}{n^2} \frac{m_e e^4}{8h^2 \varepsilon_0^2}$$

This shows that as the principal quantum number n increases, the energy also increases (becomes less negative), until at $n = \infty$ $E = 0$, and the electron is considered to be free from the influence of the nucleus.

The significance of assumption (5) given at the beginning of this Sub-section now becomes apparent. Suppose the electron of a hydrogen atom transfers from the orbit defined by $n = 2$ to the higher energy orbit defined by $n = 3$. The energy difference, denoted by ΔE, is given by

$$\Delta E = E_{(n=3)} - E_{(n=2)}$$

$$= \left[-\frac{1}{3^2} \frac{m_e e^4}{8h^2 \varepsilon_0^2} \right] - \left[-\frac{1}{2^2} \frac{m_e e^4}{8h^2 \varepsilon_0^2} \right]$$

$$= \frac{m_e e^4}{8h^2 \varepsilon_0^2} \left[\frac{1}{2^2} - \frac{1}{3^2} \right]$$

In this case the energy ΔE must have been put *into* the atom—in any form. If the atom subsequently reverts back to the lower energy state represented by $n = 2$, the same amount of energy ΔE must be emitted from the atom. Since energy can only be emitted in the form of electromagnetic radiation, i.e. as photons, a combination of Planck's law and the expression for ΔE gives

$$\Delta E = hf = \frac{hc}{\lambda}$$

$$= \frac{m_e e^4}{8h^2 \varepsilon_0^2} \left[\frac{1}{2^2} - \frac{1}{3^2} \right]$$

from which
$$\frac{1}{\lambda} = \frac{m_e e^4}{8h^3 \varepsilon_0^2 c} \left[\frac{1}{2^2} - \frac{1}{3^2} \right]$$

where λ is the wavelength of the photon emitted.

In the general case if, in a hydrogen atom, the electron reverts from the orbit defined by $n = b$ to the orbit defined by $n = a$, the above expression becomes

$$\frac{1}{\lambda} = \frac{m_e e^4}{8h^3 \varepsilon_0^2 c} \left[\frac{1}{a^2} - \frac{1}{b^2} \right]$$

Herein lies Bohr's remarkable achievement. If this expression is compared with the empirical 'Balmer–Rydberg–Ritz–Paschen' formula given in Sub-section 3.3.2 it is seen that Bohr replaced the purely experimental Rydberg constant R by the complex $\frac{m_e e^4}{8h^3 \varepsilon_0^2 c}$ which involves only known fundamental physical constants, and the evaluation of this expression gives very good agreement with the experimental value of R.

With the above model in mind the various series of spectral lines for hydrogen can be related to 'quantum transitions' of the electron from one orbit to another, and this is illustrated in *Figure*

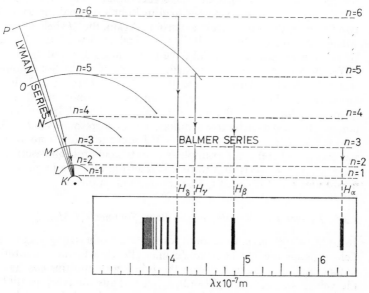

Figure 3.4. Lyman series; and Balmer quantum transitions related to spectral lines

55

3.4. The energies of the orbits defined by $n = 1, 2, 3, 4, 5$ etc. are often referred to as *K, L, M, N, O* etc. energy 'shells' as in *Figure 3.4* because, in fact, they have a finite energy 'thickness' δE rather than a single value.

3.3.4. Sommerfeld's Extension to the Bohr Model

If the spectrum of hydrogen is examined using a high resolution spectrometer it is found that each spectral line, as defined by the principal quantum number n, is made up of other, finer, lines—the spectral 'fine structure'—associated with the shell thickness δE.

The work of Sommerfeld in the field of fine structure led to the inclusion of various possible elliptical orbits representing slightly differing energies, and of an associated orbital quantum number, l, related to the eccentricity of these ellipses. For a given value of the principal quantum number n, l is limited to the values 0, 1, 2, ... $(n-1)$, which represent increasing energies, in this sequence, within the narrow energy band δE of the shell. Thus for the lowest energy, or *K*, shell (for which $n = 1$), l can only have the value 0; for the *L* shell ($n = 2$), l can have the values 0 and 1, so the *L* shell energy which was single-valued in Bohr's original theory has two values in the Sommerfeld extension. Again, the *M* shell energy, which was single-valued in Bohr's original theory, is seen to be three valued (because l can take the three values 0, 1 and 2).

The lines of the spectral fine structure, defined in the Bohr–Sommerfeld model in terms of $l = 0$, $l = 1$, $l = 2$ and $l = 3$, had already been classified by experimental spectroscopists as 'sharp', 'principal', 'diffuse' and 'fine', for which the initial letters have come to be used as definitive symbols. Thus s corresponds to $l = 0$, p to $l = 1$, d to $l = 2$ and f to $l = 3$. From the above a notation has evolved whereby, for example, an electron in an orbit described by $n = 3$ and $l = 0$ is said to be a $3s$ electron, and one in an orbit $n = 4$ and $l = 2$ is called a $4d$ electron, and so on.

3.3.5. Further Complexities in the Bohr–Sommerfeld Model

In addition to l and n, to account for the effect of a strong magnetic field on emission spectra (the Zeeman effect), a further quantum number m_l must be included in the full theory and this can have the values 0, +1, −1, +2, −2, ... $\pm l$. Thus for every possible orbit described by the orbital quantum number l there will be $(2l+1)$ values of m_l, representing $(2l+1)$ different possible energy sub-states within that orbit. For example in the orbit $n = 4$, $l = 3$ (the $4f$ state) m_l will have $(2 \times 3 + 1) = 7$ different values (i.e. -3, -2, -1, 0, $+1$, $+2$, $+3$) representing 7 different energy levels.

Finally it is observed that the spectral lines corresponding to each of the energy states represented by the above three quantum numbers, n, l, and m_l, can often be resolved into very close pairs of lines. This is explained by postulating that each orbiting electron spins about its own axis. In a particular orbit, and all other things being equal, such a spin in one sense (say clockwise) will represent a slightly different energy state from spin in the opposite sense (anti-clockwise). Both these possibilities are covered by the use of a spin quantum number m_s which can only have two values (designated $+\frac{1}{2}$ and $-\frac{1}{2}$) representing the two possible directions of spin and hence two very slightly different energy states.

The above Bohr–Sommerfeld model can be further refined by taking into account that the centre of mass of the electron–proton system is not exactly at the centre of the proton but is displaced slightly towards the electron; and also taking account of the very slight effect of relativity on the mass of the orbiting electron.

3.3.6. *Pauli's Exclusion Principle and Extra-nuclear Structure*

From the above it is seen that there are four quantum numbers n, l, m_l, and m_s, and these, taken in conjunction with Pauli's Exclusion Principle—which states 'In any atom no two electrons can have exactly the same set of four quantum numbers'— provide a complete specification for the extra-nuclear electronic structure of any atom.

This structure is indicated in the list of the natural elements contained in *Table 3.1* in which the elements are represented by their most abundant natural isotopes in their lowest energy states. Only the principal and orbital quantum states (n and l) are indicated but the number of electrons, or electron energy states, in each orbit is in accordance with the quantum laws given and with Pauli's Principle. The overlap of certain energy bands (or shells) is implied by the occupation of certain s and p states before the d and f states of the previous shell.

3.4. VALENCE ELECTRONS AND VALENCY

In the atoms of any element all those electrons in the outermost shell are *valence electrons*. In some elements several of their next inner shell d-state electrons are also valence electrons in the sense

Table 3.1

EXTRA-NUCLEAR STRUCTURE OF THE NATURAL ELEMENTS INDICATING SHELLS, AND 'BREAKDOWN' FOR THE PRINCIPAL AND THE ORBITAL QUANTUM NUMBERS n AND l.

(From *List of the Elements*, by courtesy of Mullard Ltd.)

Atomic number Z	Name of element	Chem. symbol	Atomic weight	K $n=1$	L $n=2$		M $n=3$			N $n=4$				O $n=5$				P $n=6$			Q $n=7$
				l 0 s	l 0 s	1 p	0 s	l 1 p	2 d	0 s	l 1 p	2 d	3 f	0 s	l 1 p	2 d	3 f	0 s	l 1 p	2 d	l 0 s
1	HYDROGEN	H	1.008	1																	
2	HELIUM	He	4.003	2																	
3	LITHIUM	Li	6.939	2	1																
4	BERYLLIUM	Be	9.012	2	2																
5	BORON	B	10.81	2	2	1															
6	CARBON	C	12.01	2	2	2															
7	NITROGEN	N	14.01	2	2	3															
8	OXYGEN	O	16.00	2	2	4															
9	FLUORINE	F	19.00	2	2	5															
10	NEON	Ne	20.18	2	2	6															
11	SODIUM	Na	22.99	2	2	6	1														
12	MAGNESIUM	Mg	24.31	2	2	6	2														
13	ALUMINIUM	Al	26.98	2	2	6	2	1													
14	SILICON	Si	28.09	2	2	6	2	2													
15	PHOSPHORUS	P	30.99	2	2	6	2	3													
16	SULPHUR	S	32.06	2	2	6	2	4													
17	CHLORINE	Cl	35.45	2	2	6	2	5													
18	ARGON	Ar	39.95	2	2	6	2	6													
19	POTASSIUM	K	39.10	2	2	6	2	6	—	1											
20	CALCIUM	Ca	40.08	2	2	6	2	6	—	2											

58

No.	Name	Symbol	At. wt.	1s	2s	2p	3s	3p	3d	4s	4p	4d	5s	5p
21	SCANDIUM	Sc	44.96	2	2	6	2	6	1	2				
22	TITANIUM	Ti	47.90	2	2	6	2	6	2	2				
23	VANADIUM	V	50.94	2	2	6	2	6	3	2				
24	CHROMIUM	Cr	52.00	2	2	6	2	6	5	1				
25	MANGANESE	Mn	54.94	2	2	6	2	6	5	2				
26	IRON	Fe	55.85	2	2	6	2	6	6	2				
27	COBALT	Co	58.93	2	2	6	2	6	7	2				
28	NICKEL	Ni	58.71	2	2	6	2	6	8	2				
29	COPPER	Cu	63.54	2	2	6	2	6	10	1				
30	ZINC	Zn	65.37	2	2	6	2	6	10	2				
31	GALLIUM	Ga	69.72	2	2	6	2	6	10	2	1			
32	GERMANIUM	Ge	72.59	2	2	6	2	6	10	2	2			
33	ARSENIC	As	74.92	2	2	6	2	6	10	2	3			
34	SELENIUM	Se	78.96	2	2	6	2	6	10	2	4			
35	BROMINE	Br	79.91	2	2	6	2	6	10	2	5			
36	KRYPTON	Kr	83.80	2	2	6	2	6	10	2	6			
37	RUBIDIUM	Rb	85.47	2	2	6	2	6	10	2	6		1	
38	STRONTIUM	Sr	87.62	2	2	6	2	6	10	2	6		2	
39	YTTRIUM	Y	88.91	2	2	6	2	6	10	2	6	1	2	
40	ZIRCONIUM	Zr	91.22	2	2	6	2	6	10	2	6	2	2	
41	NIOBIUM	Nb	92.91	2	2	6	2	6	10	2	6	4	1	
42	MOLYBDENUM	Mo	95.94	2	2	6	2	6	10	2	6	5	1	
43	TECHNETIUM	Tc	(97)	2	2	6	2	6	10	2	6	6	1	
44	RUTHENIUM	Ru	101.1	2	2	6	2	6	10	2	6	7	1	
45	RHODIUM	Rh	102.9	2	2	6	2	6	10	2	6	8	1	
46	PALLADIUM	Pd	106.4	2	2	6	2	6	10	2	6	10		
47	SILVER	Ag	107.9	2	2	6	2	6	10	2	6	10	1	
48	CADMIUM	Cd	112.4	2	2	6	2	6	10	2	6	10	2	
49	INDIUM	In	114.8	2	2	6	2	6	10	2	6	10	2	1
50	TIN	Sn	118.7	2	2	6	2	6	10	2	6	10	2	2

Table 3.1 (continued)

Atomic number Z	Name of element	Chem. symbol	Atomic weight	K n=1 l=0 s	L n=2 0,s	1,p	M n=3 0,s	1,p	2,d	N n=4 0,s	1,p	2,d	3,f	O n=5 0,s	1,p	2,d	3,f	P n=6 0,s	1,p	2,d	Q n=7 l=0 s
51	ANTIMONY	Sb	121.8	2	2	6	2	6	10	2	6	10	—	2	3	—	—				
52	TELLURIUM	Te	127.6	2	2	6	2	6	10	2	6	10	—	2	4	—	—				
53	IODINE	I	126.9	2	2	6	2	6	10	2	6	10	—	2	5	—	—				
54	XENON	Xe	131.3	2	2	6	2	6	10	2	6	10	—	2	6	—	—				
55	CAESIUM	Cs	132.9	2	2	6	2	6	10	2	6	10	—	2	6	—	—	1			
56	BARIUM	Ba	137.3	2	2	6	2	6	10	2	6	10	—	2	6	—	—	2			
57	LANTHANUM	La	138.9	2	2	6	2	6	10	2	6	10	—	2	6	1	—	2			
58	CERIUM	Ce	140.1	2	2	6	2	6	10	2	6	10	2	2	6	—	—	2			
59	PRASEODYMIUM	Pr	140.9	2	2	6	2	6	10	2	6	10	3	2	6	—	—	2			
60	NEODYMIUM	Nd	144.2	2	2	6	2	6	10	2	6	10	4	2	6	—	—	2			
61	PROMETHIUM	Pm	(145)	2	2	6	2	6	10	2	6	10	5	2	6	—	—	2			
62	SAMARIUM	Sm	150.4	2	2	6	2	6	10	2	6	10	6	2	6	—	—	2			
63	EUROPIUM	Eu	152.0	2	2	6	2	6	10	2	6	10	7	2	6	—	—	2			
64	GADOLINIUM	Gd	157.3	2	2	6	2	6	10	2	6	10	7	2	6	1	—	2			
65	TERBIUM	Tb	158.9	2	2	6	2	6	10	2	6	10	9	2	6	—	—	2			
66	DYSPROSIUM	Dy	162.5	2	2	6	2	6	10	2	6	10	10	2	6	—	—	2			
67	HOLMIUM	Ho	164.9	2	2	6	2	6	10	2	6	10	11	2	6	—	—	2			
68	ERBIUM	Er	167.3	2	2	6	2	6	10	2	6	10	12	2	6	—	—	2			
69	THULIUM	Tm	168.9	2	2	6	2	6	10	2	6	10	13	2	6	—	—	2			
70	YTTERBIUM	Yb	173.0	2	2	6	2	6	10	2	6	10	14	2	6	—	—	2			
71	LUTETIUM	Lu	175.0	2	2	6	2	6	10	2	6	10	14	2	6	1	—	2			
72	HAFNIUM	Hf	178.5	2	2	6	2	6	10	2	6	10	14	2	6	2	—	2			
73	TANTALUM	Ta	180.9	2	2	6	2	6	10	2	6	10	14	2	6	3	—	2			
74	TUNGSTEN	W	183.9	2	2	6	2	6	10	2	6	10	14	2	6	4	—	2			
75	RHENIUM	Re	186.2	2	2	6	2	6	10	2	6	10	14	2	6	5	—	2			

Z	Element	Symbol	At. Wt.	1s	2s	2p	3s	3p	3d	4s	4p	4d	4f	5s	5p	5d	5f	6s	6p	6d	7s
76	OSMIUM	Os	190.2	2	2	6	2	6	10	2	6	10	14	2	6	6		2			
77	IRIDIUM	Ir	192.2	2	2	6	2	6	10	2	6	10	14	2	6	7		2			
78	PLATINUM	Pt	195.1	2	2	6	2	6	10	2	6	10	14	2	6	9		1			
79	GOLD	Au	197.0	2	2	6	2	6	10	2	6	10	14	2	6	10		1			
80	MERCURY	Hg	200.6	2	2	6	2	6	10	2	6	10	14	2	6	10		2			
81	THALLIUM	Tl	204.4	2	2	6	2	6	10	2	6	10	14	2	6	10		2	1		
82	LEAD	Pb	207.2	2	2	6	2	6	10	2	6	10	14	2	6	10		2	2		
83	BISMUTH	Bi	209.0	2	2	6	2	6	10	2	6	10	14	2	6	10		2	3		
84	POLONIUM	Po	(209)	2	2	6	2	6	10	2	6	10	14	2	6	10		2	4		
85	ASTATINE	At	(210)	2	2	6	2	6	10	2	6	10	14	2	6	10		2	5		
86	RADON	Rn	(222)	2	2	6	2	6	10	2	6	10	14	2	6	10		2	6		
87	FRANCIUM	Fr	(223)	2	2	6	2	6	10	2	6	10	14	2	6	10		2	6		1
88	RADIUM	Ra	(226)	2	2	6	2	6	10	2	6	10	14	2	6	10		2	6		2
89	ACTINIUM	Ac	(227)	2	2	6	2	6	10	2	6	10	14	2	6	10		2	6	1	2
90	THORIUM	Th	232.0	2	2	6	2	6	10	2	6	10	14	2	6	10		2	6	2	2
91	PROTACTINIUM	Pa	(231)	2	2	6	2	6	10	2	6	10	14	2	6	10	2	2	6	1	2
92	URANIUM	U	238	2	2	6	2	6	10	2	6	10	14	2	6	10	3	2	6	1	2

that they contribute to valency. The outermost shell of electrons is often referred to as the *valence shell*. Reference to *Table 3.1* shows that, for example, hydrogen (H) has one valence electron, oxygen (O) has six, chlorine (Cl) seven and so on, eight being the greatest possible number of outermost shell electrons that an atom may have.

It is only valence electrons which take part in chemical reactions, and it is usually, but not always, these which are raised to higher energy states when energy is supplied to the atom. It will be seen in Chapter 6, 'Groups and Ordered Assemblies of Atoms', that valence electrons play a vital role in the structure of all materials and in their behaviour.

A study of the simple chemistry of the elements reveals that the atoms of certain of them, the *inert gases*, are very stable, that is they have practically no inclination to react in any way with other atoms. Spontaneous chemical reactions occur in response to the natural tendency of systems to lower their energy. That the inert gases do not so react is taken as evidence that their atoms are already in a low energy state and this implies a very low energy electronic structure. The electronic shell structure of these elements is shown in *Table 3.2*.

Table 3.2 ELECTRONIC SHELL STRUCTURE OF THE INERT GASES

Inert gas element		Z	$n=1$ K	$n=2$ L	$n=3$ M	$n=4$ N	$n=5$ O	$n=6$ P
Helium	(He)	2	2					
Neon	(Ne)	10	2	8				
Argon	(Ar)	18	2	8	8			
Krypton	(Kr)	36	2	8	18	8		
Xenon	(Xe)	54	2	8	18	18	8	
Radon	(Rn)	86	2	8	18	32	18	8

In each case, except helium which has an outer shell electron *duplet*, there are eight outer shell electrons, i.e. an *octet*. The valence shell octet, or K valence shell duplet, thus represents a very low energy electronic configuration. Atoms having any other configuration will have a higher energy and may be expected to seek to lower their energy by attempting to achieve the above octet (or duplet).

The *valency* of an element is the combining power of its atom; it represents the number of hydrogen atoms which an atom of the element will combine with or replace. For example the valency of oxygen in water, H_2O, is 2, and the valency of iron in the oxide FeO is 2.

The Group A elements of the chemists' periodic table—*Table 3.3*—comprise those relatively near an inert gas element in *Table 3.1*. The valency of these elements, and of the elements of Group IIB, is simply the number of their outer shell electrons, or 8 minus that number, whichever is the less.

Examples should make this clear. Sodium (Na) has valency 1, magnesium (Mg) has valency 2, oxygen (O) has valency 2, fluorine (Fl) has valency 1 and zinc (Zn) has valency 2. As examples of elements nearer to He than to Ne in *Table 3.1*, lithium (Li) has valency 1, beryllium (Be) has valency 2 and boron has valency 3.

Those elements of Groups B (except IIB) of the periodic table are further removed from an inert gas in *Table 3.1*, and have more complex electronic structures in that many have incomplete inner shell structures. Most of them exhibit more than one valency —four valencies is quite common, and manganese has five. At least one valency for these elements will be given by the number of outer shell electrons (e.g. scandium has valency 3) and/or by the number of electrons to be removed to leave an outer shell octet (e.g. tungsten exhibits valencies 2 and 6—and others; chromium exhibits a valency 6—and others).

Table 3.3. THE PERIODIC TABLE (From Moffatt, Pearsall and Wolff, *The Structure and Properties of Materials*, Vol. 1, 1964, by courtesy of John Wiley)

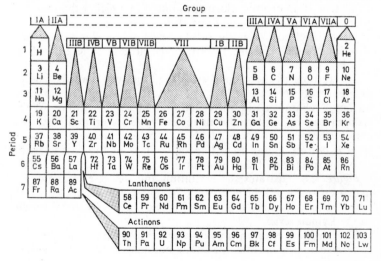

Many of the important structural elements exhibit more than one valency, for example iron (Fe) exhibits valencies 2 and 3. There is very little energy difference between its $3d$ and the $4p$ electron energy states, and whereas the lowest energy electronic pattern is

$$2, 8, 14, 2$$

as in *Table 3.1*, this can very easily assume the slightly higher energy pattern

$$2, 8, 13, 3$$

by an electron transferring from the $3d$ to the $4p$ state. The former configuration is seen to give rise to valency 2, whereas the latter suggests valency 3. In other words, in addition to the two outer shell $4s$ electrons, one of the $3d$ electrons of iron contributes to valency and is, in this sense, a valence electron.

3.5. EXCITATION, IONISATION AND IONISATION ENERGY

Table 3.1 gives the electronic configuration of single atoms in their lowest energy, or 'ground' state. In the Bohr theory for hydrogen it was shown that the orbiting electron has many possible orbits depending mainly upon the value of the principal quantum number n. If an orbiting electron is raised to a higher than normal energy state the atom is said to be *excited*. For example, a hydrogen atom whose electron is in the orbit defined by $n = 2$ is excited and, left alone, will inevitable return to the ground state defined by $n = 1$.

In returning to the ground state the atom will emit the energy difference between the $n = 2$ and the $n = 1$ states as a photon having a characteristic wavelength. Similarly an atom of any element can be excited and in returning to its ground state will emit a photon, or a series of photons, with wavelength or wavelengths corresponding to a line or to lines in the characteristic emission spectrum of that element.

Figure 3.5. illustrates the excitation and the de-excitation of a hydrogen atom and the emission of a characteristic photon, which is shown by reference to Section 3.3 and *Figure 3.4* to contribute to the Lyman series.

If an orbiting electron is completely removed from an atom this is termed *ionisation* and the atom becomes an ion. The easiest

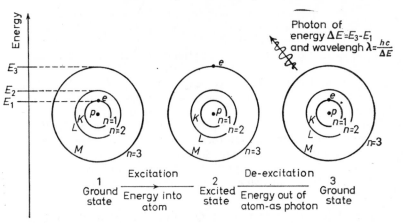

Figure 3.5. Excitation, de-excitation and emission of photon from H atom

electrons to remove from any atom are the valence electrons and the ionisation energy refers to the work necessary to completely remove a valence electron from the atom. In the case of hydrogen, using Bohr's theory, the ionisation energy is obtained by considering the electron raised from the orbit described by $n = 1$ to that described by $n = \infty$. In this case the magnitude of the ionisation energy ΔE_i is given by

$$\Delta E_i = \frac{m_e e^4}{8h^2\varepsilon_0^2}\left[\frac{1}{1^2} - \frac{1}{\infty^2}\right]$$

$$= \frac{m_e e^4}{8h^2\varepsilon_0^2}$$

$$= 2.18\times10^{-18}\ \text{J}$$

and this is about 13.6 eV.

3.6. QUESTIONS

1. What is the magnitude of the uniform magnetic flux density which allows an electrically charged particle to travel undeviated with velocity 5×10^5 m s^{-1} between, and parallel to, two parallel plates 5 mm apart with a potential difference of 2 kV between them?

65

2. An alloy steel connecting rod fails due to corrosion. A Bainbridge type mass spectrograph, with facilities for both positive and negative ion analysis, is used to examine a sample of the corrosion product (scale) in an attempt to determine the cause of failure.

The electric field in the velocity selector is 2×10^4 V m^{-1} and both the magnetic fields of the instrument are set at 0.20 T. When the instrument is set up for negative ion analysis there are two intense lines on the recorder plate. One of these is at 8.31×10^{-2} m from the ion entry slit and the other at 16.65×10^{-2} m. Assuming the ions thus recorded are doubly charged, what elements are present in the scale as represented by these lines? (When using the list of elements in *Table 3.1*, take atomic weights as masses in a.m.u.)

3. When the instrument is Question 2 is switched to analyse positive ions there are two relatively faint lines at 32.96×10^{-2} m and 28.50×10^{-2} m from the ion entry slit. What elements do these lines represent, again assuming doubly charged ions?

4. The three natural isotopes of silicon are $^{28}_{14}$Si of mass 27.976 9 a.m.u. (by mass spectrograph) and 92.21% abundance, $^{29}_{14}$Si of mass 28.976 5 a.m.u. and 4.700% abundance, and $^{30}_{14}$Si of mass 29.973 8 a.m.u. and 3.090% abundance.

Calculate the atomic weight of natural silicon.

5. Use the 'Balmer–Rydberg–Ritz–Paschen' formula to determine the wavelength of the lowest energy spectral line of the Paschen series of the hydrogen spectrum.

6. Make use of the assumptions of the Bohr theory to calculate the radius of the hydrogen atom in its lowest energy state. What is the orbital speed of the electron in this state?

7. Calculate the energy required to raise a hydrogen atom from the ground, or lowest energy, state to the state described by $n = 3$ in the simple Bohr model.

What is the wavelength of the radiation emitted when the atom reverts directly from the $n = 3$ state to the ground state?

8. Use the methods of the simple Bohr theory to determine the second ionisation energy of helium, in other words the energy to completely remove the remaining extra-nuclear electron from the helium ion $^4_2$He$^+$.

Chapter 4

The Structure and Properties of Atomic Nuclei

4.1. NUCLEAR STRUCTURE

In the Rutherford–Bohr–Sommerfeld model of the atom the nucleus consists of tightly-bound protons and neutrons, and these are collectively referred to as *nucleons*. Sometimes it is convenient to think of the neutron as comprising a proton with an embedded electron, thus being a neutral particle—this was an early concept which is strictly obsolete, but it still has some uses, especially in visualising certain aspects of radioactivity.

One of the major difficulties in accepting the above idea of nuclear structure is the following apparent paradox. Because atomic nuclei comprise tightly packed protons and neutrons there must be, by Coulomb's law, very strong electrostatic repulsive forces between the protons. These will tend to disrupt the nucleus, yet in practice the nucleus is known to be an extremely strongly bound structure, very difficult to break down. This apparent paradox will be dealt with first.

4.1.1. The Nuclear Mass Defect (D)

Atomic mass units have already been defined in Sub-section 3.2.3 and values for the masses of the electron, proton and neutron expressed in these units are

$$m_e = 0.000\ 549 \text{ a.m.u.}$$
$$m_p = 1.007\ 276 \text{ a.m.u.}$$
$$m_n = 1.008\ 665 \text{ a.m.u.}$$

From straightforward considerations, the mass of any nucleus might be expected to be the sum of the masses of its constituent nucleons. For example, consider the nucleus $^2_1$H* (the deuterium ion or 'deuteron'); its expected mass is given by

$$\text{Nuclear mass of } {}^2_1\text{H} = m_p + m_n$$
$$= 1.007\ 276 + 1.008\ 665 \text{ a.m.u.}$$
$$= 2.015\ 941 \text{ a.m.u.}$$

But direct measurement, for example using a Bainbridge mass spectrometer (*see* Sub-section 3.2.2), shows that the mass is 2.013 551 a.m.u. so that there is a loss of mass of 0.002 390 a.m.u. when the proton and neutron combine to form the nucleus of $^2_1$H. This 'lost' mass is referred to as the *nuclear mass defect* (*D*), and there is a mass defect for all nuclei with a mass number greater than 1.

4.1.2. Nuclear Binding Energy and Short-range Forces

The significance of the mass defect is contained in Einstein's mass–energy equivalence relationship:

$$E = mc^2$$

Staying with $^2_1$H as the example, when the proton and neutron combine to form this nucleus, equivalent mass is emitted in the form of energy ('fusion' energy). The numerical value of this energy can be obtained using $E = mc^2$, which in SI units gives

$$E = (0.002\ 390 \text{ a.m.u.} \times 1.660\ 4 \times 10^{-27} \text{ kg/a.m.u.})$$
$$\times (2.997\ 9 \times 10^8 \text{ m/s})^2$$
$$= 3.567 \times 10^{-13} \text{ kg m}^2 \text{ s}^{-2} \text{ (i.e. joules)}$$

and this last figure converts to 2.227 MeV (million electron volts).

---

* In this chapter, which specifically deals with nuclei, the charges are omitted from the symbolic notation except when required to avoid ambiguity.

For such calculations it is sometimes helpful to use the relationship

$$1 \text{ a.m.u.} \approx 931 \text{ MeV}$$

which is obtained from $E = mc^2$.

The emission of 2.227 MeV (3.567×10^{-13} J) on the formation of one ${}_{1}^{2}H$ nucleus implies that to split a ${}_{1}^{2}H$ nucleus into its constituent proton and neutron will need the supply of the same amount of energy. This figure 2.227 MeV, or 3.567×10^{-13} J, is referred to as the *binding energy* of the ${}_{1}^{2}H$ nucleus because it is, in effect, the energy holding the nucleus together.

The magnitude of the actual forces holding the nucleons together cannot be specified. By definition, energy is equal to force times distance, and although the magnitude of the energy is known, it has so far proved impossible to assign a value to the precise distance over which the nuclear forces apply. It is only known that these forces are effective over very short distances so they are called *short-range forces*, and within these distances they completely dominate the coulomb forces of repulsion between the protons in any nucleus.

4.2. UNSTABLE NUCLEI—RADIOACTIVITY

4.2.1. General

Certain isotopes, both naturally occurring and artificially produced, have the property of spontaneously transforming their nuclear structure. Usually a series of such transformations occurs, coming to an end when a completely stable nuclear structure is arrived at. These isotopes are termed *radioactive* because their spontaneous nuclear transformations are all energy-reducing and involve the ejection of energy, or its mass equivalent, from the nucleus. This ejected energy is termed *radiation*. Radioactive nuclear transformations, and their associated radiations, will continue irrespective of practically all physical and chemical circumstances, and in this fact lies their great usefulness in modern technology.

No account is taken here of any effects of outgoing radiation upon the extra-nuclear structure of the emitting atom, and such relatively uncommon modes of decay, and types of radiation, as *K*-capture, internal conversion, isomeric transition and neutron and proton emission are not considered.

There are three principal forms in which energy is emitted from radioactive nuclei, and because these emissions were unidentified when they were first discovered they were named as the initial three letters of the Greek alphabet. They are:

(*i*) *Alpha* (α) *emission*—the ejection from the radioactive nuclei of energetic particles which have been identified as helium nuclei ($^4_2$He). In this process a radioactive nucleus, the *parent*, emits an alpha particle and is thereby transformed into a different nucleus, the *daughter*.

For example, the natural radium isotope $^{226}_{88}$Ra is an alpha emitter and it produces the daughter ($^{222}_{86}$) which is an isotope of the element radon. This emission process may be written symbolically as follows (*see* footnote to Sub-section 4.1.1)

$$(^{226}_{88}) \rightarrow (^{222}_{86}) + (^4_2)$$

or, more usually

$$(^{226}_{88}) \rightarrow (^{222}_{86}) + \alpha$$

Alpha emission only occurs from the heavy elements, i.e. those having an atomic number greater than 77.

(*ii*) *Beta* (β) *emission*—the ejection of particles which have been identified as electrons (β^-) in some cases and as positrons (β^+) in others, (*see also* Section 2.4). As with alpha radiation, the emission of beta particles transforms the parent nucleus into a daughter which is a different element. For example the carbon isotope ($^{14}_6$) is a β^- emitter and, using the mental picture of a neutron comprising a proton with an embedded electron (*see* Section 4.1), the daughter nucleus can be deduced to be ($^{14}_7$), a stable isotope of the element nitrogen. Another isotope of carbon ($^{10}_6$) is a β^+ emitter and in this case the removal of a positive charge from the nucleus effectively converts a proton into a neutron. The daughter nucleus is thus ($^{10}_5$), a stable isotope of the element boron.

The above two examples can be expressed symbolically as

$$(^{14}_6) \rightarrow (^{14}_7) + \beta^-$$

and

$$(^{10}_6) \rightarrow (^{10}_5) + \beta^+$$

The emission of β^- is the most common form of radiation.

(*iii*) *Gamma* (γ) *emission*—the ejection of pure energy, as photons. Gamma photons have very high energies and therefore very short wavelengths $\left(\text{remember } E_q = hf = \dfrac{hc}{\lambda}\right)$. The emission of gamma very rarely occurs on its own, being almost invariably associated with the prior emission of beta or, less commonly, alpha. In such cases the emission of the beta or alpha produces a daughter which is in an excited, or elevated energy, state and this energy is reduced to the lowest energy, or ground, state by the emission of the gamma photon. A typical reaction of this kind is shown symbolically thus

$$\left(^{220}_{86}\right) \rightarrow \left(^{216}_{84}\right) + \alpha$$
$$\text{Excited}$$
$$\downarrow$$
$$\left(^{216}_{84}\right) + \gamma$$
$$\text{Ground}$$

or simply as

$$\left(^{220}_{86}\right) \rightarrow \left(^{216}_{84}\right) + \alpha + \gamma$$

The excited state is usually extremely short-lived.

4.2.2. Radioactive Decay—Half-life and Activity

When a nucleus emits radiation and transforms into a daughter nucleus the parent is said to have decayed. Radioactive decay is therefore the process by which a mass of a particular radioactive isotope more or less slowly transforms into other substances (daughters, granddaughters, great granddaughters etc.).

The mathematical law describing radioactive decay is a statistical population law stating that at any instant the rate of decay of a radioactive isotope is proportional to the quantity of that isotope present in the sample.

Suppose N is the number of atoms of a particular isotope in a sample, and t represents time, then the above may be written

$$\frac{dN}{dt} \propto N$$

(This law is analogous to the population law 'the greater the population of a particular living species the greater will be the total number of deaths per unit time—other things being equal'.)

Putting a constant of proportionality, λ, into the above gives

$$-\frac{dN}{dt} = \lambda N$$

The minus sign merely indicates that N decreases with time, and λ is called the *decay constant* or *transformation constant*.

71

If N_0 is the number of parent atoms in the sample at time $t = 0$ and N_t is their number at some later time t, integration of this simple differential equation is as follows

$$\int_{N_0}^{N_t} \frac{dN}{N} = -\int_0^t \lambda \, dt$$

whence

$$\log_e \frac{N_t}{N_0} = -\lambda t$$

or

$$N_t = N_0 e^{-\lambda t}$$

In other words the decay law is exponential, but this is only valid for large values of N (not a severe restriction when one considers that the number of atoms per mole of any element is of the order* of 10^{23}).

A graph of N_t against t is shown in *Figure 4.1* and indicates that N_t never becomes zero but only tends to zero as t tends to infinity. Because of this it is not possible to talk of the 'life' of a sample of a radioactive isotope in the sense of the time for N_t to become zero. However there is a significant factor that can be used with an exponential function, and this is the 'half-value'. In the case of radioactive decay the half-value of interest is the *half-life* (symbol $t_{1/2}$), and this is the time for N to reduce to half its value. Half-life is sometimes called the *period* of an isotope, and its significance is shown in *Figure 4.1*.

Mathematically the half-life or period is easily obtained if the transformation constant (λ) is known, thus

from

$$N_t = N_0 e^{-\lambda t}$$

$$\frac{N_0}{2} = N_0 e^{-\lambda t_{1/2}} \qquad \text{[by definition of half-life]}$$

$\therefore$

$$\tfrac{1}{2} = e^{-\lambda t_{1/2}}$$

$$2 = e^{+\lambda t_{1/2}}$$

$$t_{1/2} = \frac{\log_e 2}{\lambda}$$

$$t_{1/2} \approx \frac{0.693}{\lambda}$$

---

* The Avogadro constant (N_A) is $6.022\,52 \times 10^{23}$ and is by definition the number of elementary units per mole of a substance. *See also* Sub-section 1.1.4.

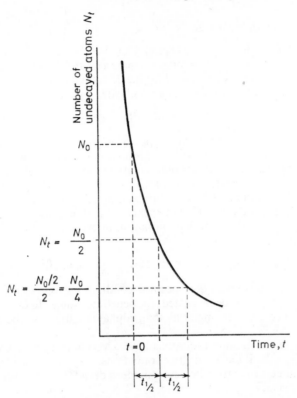

Figure 4.1. Concept of half-life ($t_{1/2}$) in radioactive decay. (N. B. In place of N_t it is common to plot dN/dt. This is acceptable because $dN/dt \propto N$)

The *activity* of a sample of radioactive isotope is the number of transformations which occur per unit time, and is therefore $-\dfrac{dN}{dt}$.
The strict SI unit of activity is the hertz (Hz) or s^{-1}, that is the actual number of transformations per second. The traditional unit the *curie* (Ci) is still very widely used and is defined to be that quantity of a radioactive isotope which produces 3.7×10^{10} transformations per second. This unlikely-seeming number originates from an early determination of the *specific activity*, that is the activity per gramme, of radium, which was taken as the standard.

4.2.3. The Detection of Radiation

All three radiations, alpha, beta and gamma, possess the property of ionising some of the atoms of materials through which they travel. This is the property used for the detection of such radiations. Their relative ionising abilities are indicated by the ratios

$$\text{alpha, } 10^4: \qquad \text{beta, } 10^2: \qquad \text{gamma, } 10^0$$

which are to be taken as very approximate orders of magnitude only. In this context, the term *specific ionisation* is more appropriate and is defined as the number of ion pairs [positive ion + electron(s)] produced per particle, or photon, per unit path length under specified conditions of temperature and pressure.

The ability of the radiations to penetrate matter is about inversely proportional to their specific ionisation, in other words it is in the approximate ratios

$$\text{alpha, } 10^0: \qquad \text{beta, } 10^2: \qquad \text{gamma, } 10^4$$

These indicate that under specified conditions gamma radiation, for example, can be expected to penetrate some 10 000 times further into a given material than alpha radiation of the same energy.

Because the gamma radiations have a very much lower specific ionisation than the other two radiations they are considerably more difficult to detect; however, modern detectors have overcome this problem.

One of the main detectors for radiation is the Geiger-Müller tube. A form of this, shown in *Figure 4.2*, comprises a gas filled metal chamber with a central electrode and a thin 'window' of mica at one end. Individual alpha or beta particles or gamma photons entering the tube chamber each produce some ionisation of the gas. A potential difference of several hundred volts between the metal case and the central electrode causes electrons freed by this ionisation to migrate rapidly to the central electrode, producing further ionisation by collision on the way. The positive ions migrate to the outer case, but more slowly because of their greater mass.

This charge separation results in a pulse of electrons through R, the positive ions being discharged by collecting electrons at the metal case. The electronic pulse across R can be recorded in many ways, for example by audible, visual or graphical methods. The Geiger tube is usually connected either to a *scaler*, which is a pulse

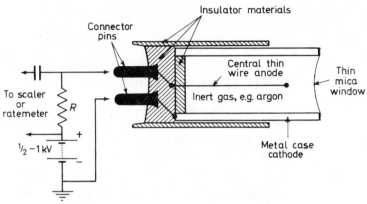

Figure 4.2. Essentials of Geiger-Müller tube

counter giving a direct visual record of the actual number of pulses, or to a *ratemeter* which gives direct visual indication of the pulse rate, in other words the number of pulses per unit time. This last is, of course, a direct measure of the activity $\dfrac{dN}{dt}$.

4.2.4. Some Uses of Radio-isotopes

In recent years very rapid advances have been made in the understanding and production of radio-isotopes. By placing certain materials in a nuclear reactor, in which they are subjected to a flux of neutrons, or by subjecting them to bombardment by energetic particles (for example protons or deuterons) from a particle accelerator such as a cyclotron, nuclear transformations can be effected. These transformations can produce desirable radioactive isotopes of the materials concerned.

Small quantities of such active material may be added to the natural, non-active, material used in any industrial, medical or research process. The radiation from the radioactive material enables its progress to be traced throughout the process by using suitable detector and ancillary equipment. Used in this way radio-isotopes are termed *tracers*, and today their use is commonplace.

An interesting example of the use of tracer techniques is the determination of the location of very small percentage additions of elements in alloys. For example a small proportion of a beta-active isotope of the element concerned could be added to the

75

molten metal in the usual way. A test sample of the alloy would be cast and prepared for examination. The fact that beta radiation will expose a photo-emulsion enables a contact radiograph to be taken. In conjunction with microscopic examination, this will show the location of the active material in relation to the grain structure of the alloy and hence the location of the alloying element.

The penetrative and absorptive properties, particularly of beta radiation, can be put to many industrial uses. For example, radio-isotopes are used in quality control such as automatic 'on stream' thickness monitoring of a product manufactured in sheet form. The count rate from a beta source which is placed on one side of the material is obtained by a detector on the other side, and this count rate will depend upon the amount of radiation absorption in the material, which in turn depends upon the thickness of the material. Suitable 'feedback' from the counting device can enable the thickness control to be fully automated. In a process involving unit production a similar feedback could operate a rejection mechanism for units whose thickness was outside the manufacturers' tolerance limits.

The same absorption properties enable beta- (and sometimes gamma-) emitting isotopes to be used for the detection—and automatic rejection—of insufficiently filled containers of a packaged product on an assembly line. Many other examples exist.

Gamma radiation is widely used in industrial radiography as an extension of the technique of x-ray photography. The properties of greater penetration and ability to expose photo-emulsions enable gamma-ray shadow photographs to be taken of such subjects as metal castings and welds, which can then be examined for flaws. Such industrial radiography is one of the most important techniques in the modern technology of non-destructive testing (N.D.T.).

4.2.5. Radiation Hazards and Shielding

Damage to living (including human) cells is caused by the ionising effects of radiation on the tissues. Alpha radiation is not very penetrating but is highly ionising and the main risk to be avoided is direct contamination (physical contact) with the alpha-emitting substance, particularly by way of inhalation and ingestion. It is comparatively easy to provide shielding against alpha radiation —even a thin piece of paper will often suffice—because of its low penetration.

Beta radiation is more penetrating, but less ionising, than alpha. It can cause internal burns and requires much more substantial shielding than alpha radiation. Lead shields and thick lead-containing glass are frequently used. One of the problems of radiation burns is that, unlike physical contact with a very hot object which causes an involuntary reflex withdrawal, the effects of even serious radiation burns are not immediately felt and there is no rapid physiological, protective, response. Any one who has suffered burns due to over-long sun bathing—a mild form of radiation damage—will know about, and appreciate the dangers of, this delayed response.

Gamma radiations are the most penetrating and therefore require the greatest shielding precautions. For example, the radiations from some gamma-emitting isotopes will easily penetrate a 30 cm thick piece of steel without any great loss in intensity; and gamma radiations will produce similar deep, delayed sensitivity, burns to those produced by beta radiations.

The subjects of radiation hazards and shielding are extensive and complex and include a host of safety regulations governing the use of radio-isotopes. Further consideration of these is outside the scope of this book.

4.3 NUCLEAR REACTIONS

4.3.1. General

Although nuclear reactions may not, at first sight, appear to belong in a study of Materials Science they are in fact highly relevant. For example the present state of knowledge of the generation of nuclear energy derives directly from a deep understanding of the nature of the materials of the 'fuel'. It should be realised that it was known for a long time before the practical achievement of the release of nuclear energy that such energy was available. The great problems were the practical ones of releasing it and then harnessing it. In other words the development of modern nuclear power technology progressed from a thorough knowledge of the theory of the atomic, and particularly the nuclear, structure of materials towards the practical release of energy, and then to its subsequent control and utilisation.

Nuclei can be bombarded with energetic projectiles—particles such as protons (p or ^1_1H), neutrons (n or ^1_0n) or alphas (α or ^4_2He). The most interesting reactions of the target nuclei to such bombardment are nuclear transformations. These events can be represented by symbolic 'equations' in which the total energy on the left hand side balances that on the right hand side, and the total atomic numbers and mass numbers on the two sides also balance.

A concise notation for nuclear reactions is

$$\text{Target nucleus} \left(\begin{array}{c} \text{Incoming particle(s)} \\ \text{or quanta} \end{array} , \begin{array}{c} \text{Outgoing particle(s)} \\ \text{or quanta} \end{array} \right) \text{Product nucleus}$$

In a nuclear reaction it is assumed that the target nucleus captures the incident particle, giving rise to a short-lived 'compound' nucleus. This immediately does any one of three things.

(*i*) Being stable in form, but excited, it emits photons to reduce its energy, and retains its identity.

Example:

$$\underset{\text{(Projectile)}}{{}^{1}_{1}\text{H}} + \underset{\text{(Target)}}{{}^{27}_{13}\text{Al}} \rightarrow \underset{\substack{\text{(Compound} \\ \text{nucleus)}}}{({}^{28}_{14}\text{Si})} \rightarrow \underset{\text{(Product)}}{{}^{28}_{14}\text{Si}} + \underset{\text{(Photon)}}{\gamma}$$

Or, written in the more concise notation

$${}^{27}_{13}\text{Al}(p, \gamma){}^{28}_{14}\text{Si}$$

but this is less instructive than the fuller equation above.

(*ii*) It becomes radioactive.

Example:

$${}^{1}_{1}\text{H} + {}^{12}_{6}\text{C} \rightarrow ({}^{13}_{7}\text{N}) \rightarrow {}^{13}_{6}\text{C} + \beta^{+} + {}^{0}_{0}\nu + E_{R}$$

Or $${}^{12}_{6}\text{C}(p, \beta^{+}\nu){}^{13}_{6}\text{C}$$

${}^{0}_{0}\nu$ or ν is a neutrino (*see* Section 2.4) and E_{R} is the reaction energy which is shared by the β^{+} and ${}^{0}_{0}\nu$ and is the energy equivalent of the mass difference between the L.H.S. and R.H.S. of the above. The kinetic energies of the projectile ${}^{1}_{1}\text{H}$ and of the product nucleus can usually be ignored.

(*iii*) Being stable in form but excited, it emits particles and transforms to some other isotope.

Example:

$${}^{4}_{2}\text{He} + {}^{27}_{13}\text{Al} \rightarrow ({}^{31}_{15}\text{P}) \rightarrow {}^{30}_{14}\text{Si} + {}^{1}_{1}\text{H} + E_{R}$$

Or $${}^{27}_{13}\text{Al}(\alpha, p){}^{30}_{14}\text{Si}$$

$^{31}_{15}$P is normally a stable isotope, but in this case its energy of excitation causes ejection of the proton $^1_1$H (or p). The reaction energy E_R possessed by the proton is again the equivalent of the mass difference between the L.H.S. and R.H.S. of the equation.

The results of nuclear bombardment are not unique. For example, instead of the result shown in *Example* (*iii*) the following could have occurred:

$$^4_2He + ^{27}_{13}Al \rightarrow (^{31}_{15}P) \rightarrow ^{30}_{15}P + ^1_0n + E_R$$

Or

$$^{27}_{13}Al(\alpha, n)^{30}_{15}P$$

4.3.2. Nuclear Energy—Fission and Fusion

Having grasped the significance of mass defect and nuclear binding energy (Sub-sections 4.1.1 and 4.1.2) it is possible to present mass defect data in a form that enables the source of nuclear energy to be visualised quite easily.

If the real mass M of the nucleus of an isotope ^A_ZX is determined using a mass spectrometer such as Bainbridge's*, the mass defect, D, is given by

$$D = Zm_p + (A-Z)m_n - M$$

and if D is in a.m.u., the nuclear binding energy (B.E.) is given, in MeV, by

$$\text{B.E.} \approx 931D \quad [\textit{see} \text{ Sub-section 4.1.2}]$$

To make a comparison of the nuclear binding energies of different elements, the quantity used must be the *specific binding energy*, S.B.E., which is the binding energy per nucleon. This is given in MeV by

$$\text{S.B.E.} \approx \frac{931D}{A}$$

This quantity S.B.E. can be thought of as the average value of the energy necessary to remove one nucleon from a multi-nucleon nucleus.

It should be clear that the greater the binding energy of any nucleus the greater was the amount of energy *emitted* when the nucleus was first formed from its constituent prime particles. So, the greater the specific binding energy, the greater will be the

---

* To obtain the nuclear mass, the mass of the extranuclear electrons of the ion is subtracted from the mass of the ion observed using the spectrometer.

energy emitted per nucleon on the formation of the nucleus. *An increase in the* S.B.E. *of a system must therefore be accompanied by the emission of energy.*

The work of Aston and later mass spectroscopists has enabled a graph of S.B.E. against mass number *A* to be plotted and this has the form shown in *Figure 4.3*.

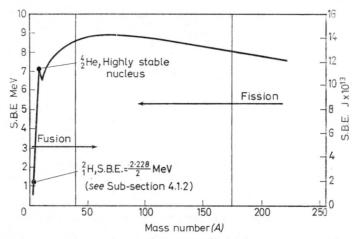

Figure 4.3. Specific nuclear binding energy (S.B.E.) and concepts of fission and fusion

This graph shows an increase in S.B.E. in moving from the high atomic mass region marked 'Fission' into the central region of moderate atomic mass. Fission refers to the splitting of a heavy nucleus, from the region so marked, into two moderately massive parts which are each constituents of the central region of high S.B.E. Fission thus involves an *increase* in S.B.E., and since there is no change in the total number of nucleons in the system, there must be an *emission* of energy.

Such a reaction is only useful if it is self-sustaining and this can be achieved in several reactions, of which the following is an example.

$$\underset{\text{(Projectile)}}{^{1}_{0}\text{n}} + \underset{\substack{\text{(Fissionable}\\\text{nucleus of}\\\text{'fuel')}}}{^{235}_{92}\text{U}} \rightarrow (^{236}_{92}\text{U}) \rightarrow \underset{\substack{\text{(Moderate mass}\\\text{product nuclei)}}}{^{94}_{38}\text{Sr} + ^{140}_{54}\text{Xe}} + 2\,^{1}_{0}\text{n} + E_{R}$$

Or $^{235}_{92}\text{U}(\text{n, 2n})^{94}_{38}\text{Sr, }^{140}_{54}\text{Xe}$

The projectile is a low energy (slow) neutron and the reaction is seen to produce two neutrons. These two, if appropriately slowed down (in a *moderator*), can both produce a similar reaction, and these in turn will each produce two neutrons and so on, giving a *chain reaction* assuming an adequate supply of fuel.

In a so-called atomic bomb, the chain reaction is uncontrolled and goes on to completion*. In an atomic reactor (the heart of a nuclear power plant) the reaction rate is held at a suitable level by using neutron-absorbing control rods. A chain reaction is represented schematically in *Figure 4.4*.

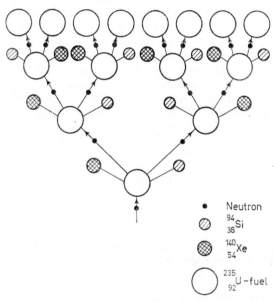

Neutron •

$^{94}_{38}Si$ ⊘

$^{140}_{54}Xe$ ⊛

$^{235}_{92}U$ – fuel ◯

Figure 4.4. Schematic representation of three steps of a typical chain reaction

Fission energy has, of course, been available in both uncontrolled form (the atomic bomb) and controlled form (nuclear power plant) for many years now and is comparatively commonplace.

Fusion energy involves moving a system of light nuclei from the region so marked in *Figure 4.3* towards the central region of near-maximum S.B.E. This necessitates causing light nuclei to combine

---

* Strictly, the reaction goes on until the explosive energy completely disperses the remaining fuel.

(fuse) and produce heavier nuclei, so creating an increase in the S.B.E. of the system. This can be done by bringing into very close proximity nuclei of such material as hydrogen ($_1^1H$) or deuterium ($_1^2H$) to create, for example, tritium ($_1^3H$) and helium ($_2^4He$) with consequent increase in S.B.E. and emission of energy.

Fusion energy has been successfully released many times in an uncontrolled form (the so-called hydrogen bomb), but the technological ambition of harnessing the energy of fusion reactions has not, at the time of writing,* been achieved. The difficulties are very great and much research is in progress.

To achieve fusion, before the short-range nuclear forces of attraction can come into play the coulomb repulsive forces between protons have to be overcome and these forces are very great when the particles get very close. For an uncontrolled fusion reaction these coulomb repulsive forces can be overcome by using the great energy of nuclear fission. A fission explosion is arranged to bring the fusion fuel nuclei sufficiently close together to enable the short-range forces to take over and cause the release of fusion energy.

An understanding of the source of fusion energy will be assisted by considering the following possible reactions which occur in rapid succession, involving deuterium ($_1^2H$) as fuel. It is assumed that there is plenty of fuel available.

$$_1^2H + _1^2H \rightarrow _1^3H + _1^1H + E_1 \qquad \ldots(4.1)$$
$$+$$
$$_1^2H \rightarrow _2^4He + _0^1n + E_2 \qquad \ldots(4.2)$$

Reaction (4.2) follows directly upon (4.1), and both depend upon the input of an appropriate amount of fission energy. The following nuclear masses are known from mass spectrometry,

$_1^1H$ 1.007 276 a.m.u.

$_1^2H$ 2.013 553 a.m.u.

$_1^3H$ 3.015 500 a.m.u.

$_0^1n$ 1.008 665 a.m.u.

$_2^4He$ 4.001 506 a.m.u.

Taking the masses on each side of the reactions and disregarding the input of fission energy, the differences between the left hand sides and right hand sides will give the mass equivalents of the energies E_1 and E_2.

* September 1970.

Reaction (4.1):

L.H.S.	2.013 553	R.H.S.	1.007 276
	+ 2.013 553		+ 3.015 500
	4.027 106		4.022 776

$$E_1 = 0.004\ 330 \text{ a.m.u.}$$

Reaction (4.2):

L.H.S.	3.015 500	R.H.S.	1.008 665
	+ 2.013 553		+ 4.001 506
	5.029 053		5.010 171

$$E_2 = 0.018\ 882 \text{ a.m.u.}$$

$$E_1 + E_2 = 0.023\ 212 \text{ a.m.u.}$$

and since 1 a.m.u. is approximately equivalent to 931 MeV, and in Reactions (4.1.) and (4.2) three atoms of fuel (^2_1H) are used, the amount of fusion energy released per atom of fuel consumed is

$$\approx \frac{931 \times 0.023\ 212}{3} \text{ MeV}$$

$$\approx 7.20 \text{ MeV}$$

$$\approx 1.15 \times 10^{-12} \text{ J}$$

This, of course, represents an enormous amount of energy per unit mass of fuel (about 3.5×10^{14} J kg^{-1}) and everyone today is fully aware of the immense energy potential of fusion reactions, having read only too often of the devastating effects of the 'H' bomb.

4.4. QUESTIONS

1. The atomic mass of the oxygen isotope $^{16}_8\text{O}$ is 15.994 91 a.m.u. as determined using a mass spectrograph. Using the known values of the electron, proton and neutron masses, make a determination of the nuclear mass defect (D) of this isotope in SI units.

2. What is (*a*) the binding energy of the nucleus of $^{16}_8\text{O}$ and (*b*) its specific binding energy?

3. The isotope $^{218}_{84}\text{Po}$ emits an α particle to become ^A_ZX which in turn emits a β^- particle and a γ photon to become $^{A'}_{Z'}\text{Y}$ and this is

part of the uranium radioactive series. What are the values of Z, A, Z', and A'? Also name the elements X and Y.

4. The half-life of radium-226 is 1600 years. If a hospital purchases one gramme of this isotope (*a*) how much of it will be left in eight hundred years? (*b*) What will be the activity of this isotope then?

5. For the nuclear fission reaction

$$^{235}_{92}U(n, 2n)\,^{94}_{38}Sr,\,^{140}_{54}Xe$$

calculate a value for the reaction energy (disregard the kinetic energy of the incident neutron). Use the following atomic masses:

$$^{235}_{92}U = 235.043\ 9 \text{ a.m.u.} \qquad ^{140}_{54}Xe = 139.912\ 6 \text{ a.m.u.}$$
$$^{94}_{38}Sr = 93.876\ 9 \text{ a.m.u.}$$

6. For the nuclear reaction of Question 5, how much energy would be emitted during the consumption by fission of 1 kg of the fuel $^{235}_{92}U$?

7. For the fusion reaction

$$^{2}_{1}H + ^{2}_{1}H \rightarrow ^{1}_{0}n + ^{3}_{2}He$$

use the following atomic masses to determine the reaction energy, disregarding the energy necessary to bring the fuel nuclei sufficiently close for fusion to occur:

$$^{2}_{1}H = 2.014\ 102 \text{ a.m.u.} \qquad ^{3}_{2}He = 3.016\ 030 \text{ a.m.u.}$$
$$^{1}_{0}n = 1.008\ 665 \text{ a.m.u.}$$

Chapter 5

Introduction to X-radiology

5.1. THE NATURE AND SOURCE OF X-RAYS

5.1.1. General

X-radiation is a high energy phenomenon and is emitted from a material when sudden, large, reductions of electron energy occur in the material. X-rays comprise streams of high energy photons with wavelengths in the electromagnetic waveband intermediate between the gamma and the ultra-violet radiation wavebands. The higher energy x-rays are termed *hard* whilst those of lower energy are termed *soft*. The name x-radiation was given because when they were first observed their nature was unknown.

Excitation of the Rutherford–Bohr–Sommerfeld atom consists of raising its energy by using any means to transfer an electron (most commonly a valence electron) to a higher energy orbit. De-excitation is the reverse process, during which the energy difference between the high energy and low energy states is emitted as photons.

Characteristic radiations in the x-ray waveband are not obtained from the lightest elements (atomic numbers less than 11 say) because even the largest possible single changes of electron energy in atoms of these materials are insufficient to produce photon wavelengths in the defined x-ray waveband—roughly $\lambda = 10^{-9}$ m to $\lambda = 10^{-12}$ m (i.e. photon energies of about 10^3 to 10^6 eV or around 10^{-16} to 10^{-13} J). For example, the maximum energy photon emitted by a hydrogen atom is only of the order of 10 eV ($\lambda \approx 10^{-7}$ m, which is in the ultra-violet), and that for sodium ($Z = 11$) has a wavelength of about 10^{-9} m (energy of about 10^3 eV, or around 10^{-16} J) which is only just within the normally accepted x-ray waveband.

5.1.2. White X-radiation or Bremsstrahlung

If a stream of high energy (that is high velocity) electrons is directed onto a target comprising a solid piece of any suitable element ($Z > 10$) as in *Figure 5.1*, many of the electrons are caused to

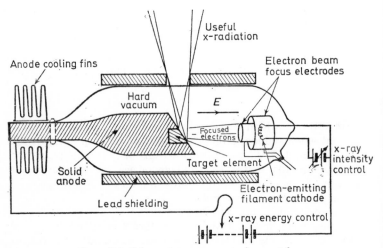

Figure 5.1. Schematic arrangement of x-ray tube

decelerate, or brake, in the vicinity of atoms of the target element, thereby losing energy. Such decelerating events can be thought of as collisions. The difference between the energy of an electron before and after such a deceleration, ΔE, is emitted as a photon of wavelength λ where

$$\lambda = \frac{hc}{\Delta E} \qquad \text{[using Planck's law]}$$

and *Figure 5.2* illustrates such a decelerating event.

86

It can be inferred that ΔE may have any value within a range from zero (associated with the electron suffering no deceleration) to a finite maximum value (associated with electrons which are completely stopped, thus giving up their entire initial kinetic energy in one decelerating event). Most commonly electrons will suffer a series of decelerations each producing an appropriate photon.

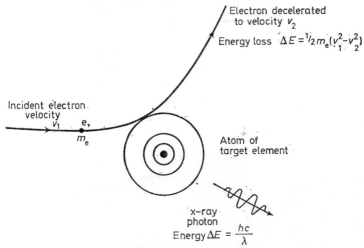

Figure 5.2. Electron deceleration and emission of photon

The photons emitted by such electron decelerations give rise to the *white* x-radiation or *bremsstrahlung* (literally 'braking radiation'). The term white radiation stems from the analogy with white light which also has a continuous spectrum, but, of course, light has no short wave (high energy) limiting value. *Figure 5.3* shows the distribution of the number of x-ray photons emitted per unit time, I (intensity), with radiation wavelength, and how this distribution is altered by varying the intensity of the incident electron beam, and the energy of the electron beam.

A great deal of the energy given up by the incident electrons is absorbed by the target material and manifests itself as heat, only a very small percentage (perhaps 0.5%) being emitted as useful x-radiation. For this reason, heat dissipation in x-ray tubes is of major concern.

87

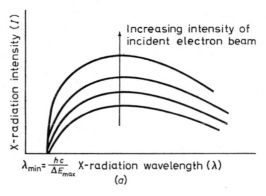

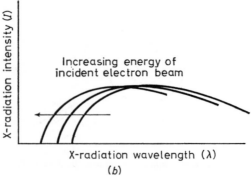

Figure 5.3. Variation of white-radiation spectrum with
(a) intensity (see Intensity control, Figure 5.1) and
(b) energy of incident electrons (see Energy control, Figure 5.1)

5.1.3. Characteristic X-radiation

If, for a given target element, the intensity of the incident electron beam is increased beyond a certain level, the emitted x-radiation has several intensity peaks at characteristic wavelengths. Such peaks are illustrated in *Figure 5.4*.

Emission at these characteristic wavelengths is due to quantum transitions by extra-nuclear electrons of the type already dealt with in discussing the hydrogen spectrum (*see* Sub-section 3.3.3).

A sufficiently energetic electron (from the incident beam) can cause the ejection of an extra-nuclear electron from one of the inner shells of a target atom. Electron transitions within this atom

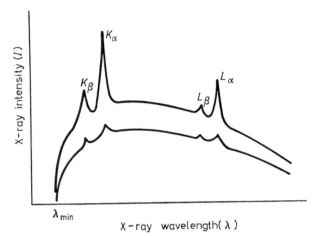

Figure 5.4. Characteristic x-ray line spectra

will then occur to fill the space in the inner shell and each transition will give rise to a characteristic photon.

Consider a target element having electrons in the four shells K, L, M and N (it could be iron). If a K shell electron is ejected by the incident electron beam its place will most probably be filled by an electron dropping down from the L shell ($L \rightarrow K$ transition), this might be followed by an $M \rightarrow L$ transition, and so on. A somewhat less probable transition to fill the K shell would be from M shell to K shell, and an even less probable one would be from N to K. $L \rightarrow K$ and $M \rightarrow K$ transitions are illustrated in *Figure 5.5*.

The $L \rightarrow K$ transition gives rise to a photon of specific energy and therefore of characteristic wavelength for the particular target element. This photon contributes to the characteristic spectral line called the K_α line. Similarly $M \rightarrow K$ transitions give rise to the K_β spectral line. Because the $L \rightarrow K$ transition is generally a more probable event than the $M \rightarrow K$ transition, the K_α line is usually of greater intensity than the K_β, but note that the K_β photons are of greater energy (shorter wavelength) than those of the K_α line.

There is also an L series of characteristic x-ray lines representing transitions to the L shell, such as $M \rightarrow L$, $N \rightarrow L$, etc., and an M series shows up for targets of the heavier elements having atomic numbers greater than about 50.

The wavelengths of the x-ray spectral lines are quite specific, being characteristic of the target element concerned. They can,

89

in fact, be used to identify even small quantities of elements contained in the x-ray target, and this principle is widely used for the physical analysis of very small samples of materials in the *electron probe micro-analyser*.

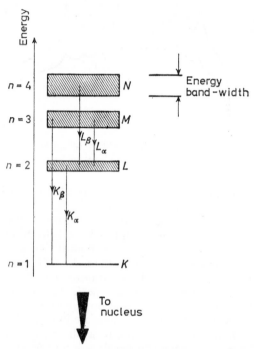

Figure 5.5. Quantum transitions for characteristic K and L x-ray lines

The extra-nuclear electron shells have finite energy **band** widths (*see Figure 5.5*) and these are known to comprise very close, but nevertheless discrete, energy levels (see also the discussion of spectral fine-structure in Sub-section 3.3.4). Because of this the characteristic x-ray spectral lines, suitably resolved, are found to comprise a fine structure. In practice the K_α and K_β lines resolve into pairs, but the L series resolves into a more complex fine structure. As a practical illustration of this fine structure the characteristic K_α line of copper has wavelength 1.54×10^{-10} m to three significant figures, but if wavelengths are measured to five significant figures the K_α line is seen to be made up of a pair of lines, $K_{\alpha 1}$ of wavelength $1.544\ 3 \times 10^{-10}$ m and $K_{\alpha 2}$ of wavelength $1.540\ 5 \times 10^{-10}$ m.

5.2. THE ABSORPTION OF X-RADIATION

5.2.1. General

White radiation is sometimes used in x-ray technology (radiology), but for many purposes monochromatic radiation (i.e. having a single wavelength, or more strictly a single narrow waveband) is required. There is little problem in obtaining white radiation because practically any target material will provide this. If only a single wavelength is required it is necessary to select a target element with the appropriate characteristic spectral line (usually a *K* line, to provide sufficient intensity for a given level of background white radiation), and then to remove, as far as possible, all other wavelengths. The removal of unwanted wavelengths is accomplished by using *filters* which will selectively transmit the required wavelength whilst largely absorbing other wavelengths.

Before discussing x-ray filters it is relevant to deal with the mathematics of absorption of monochromatic x-radiation, and this treatment is equally applicable to the absorption of monochromatic gamma radiation.

As any high energy electromagnetic radiation (x or gamma) passes through matter there is a progressive reduction in its intensity. This absorption is due partly to Compton scattering, partly to photo-electric energy transfer and, in the case of sufficiently energetic gamma ray photons, partly to direct energy-to-mass transformations. Details of these physical processes are of little concern here except that in each case part, or all, of its energy is removed from the x or gamma photon involved.

In the case of monochromatic radiation the change in intensity, I, with distance, x, travelled in an absorbing medium is $-\dfrac{\mathrm{d}I}{\mathrm{d}x}$ and is proportional to the intensity I. In other words the more photons that are present, the greater will be the number of absorbing events taking place in a given distance. This statistical law may be written

$$-\frac{\mathrm{d}I}{\mathrm{d}x} \propto I$$

91

Putting in μ as the constant of proportionality gives

$$\frac{\mathrm{d}I}{\mathrm{d}x} = -\mu I$$

and μ is called the *linear absorption coefficient*.

If I_0 is associated with $x = 0$, and I_x with distance x integration provides the following

$$\int_{I_0}^{I_z} \frac{\mathrm{d}I}{I} = -\mu \int_0^x \mathrm{d}x$$

whence

$$\log_e \frac{I_x}{I_0} = -\mu x$$

or

$$I_x = I_0 e^{-\mu x}$$

The *half-thickness* of an absorbing material for radiation of a particular wavelength is that thickness which will reduce the radiation intensity at that wavelength by half. If $x_{1/2}$ be the half-thickness then

$$\frac{I_0}{2} = I_0 e^{-\mu x_{1/2}}$$

$$2 = e^{+\mu x_{1/2}}$$

$$x_{1/2} = \frac{\log_e 2}{\mu}$$

$$x_{1/2} \approx \frac{0.693}{\mu}$$

Radiation absorbers are usually classified in terms of their *equivalent thickness* m_a which is their mass per unit area. This quantity is used in conjunction with the *mass absorption coefficient*, μ_m, which is defined as the linear absorption coefficient for the waveband under consideration divided by the density of the absorbing material. The expression

$$I_x = I_0 e^{-\mu x}$$

may be written

$$I_x = I_0 e^{-[(\mu/\varrho)\varrho x]}$$

where ϱ is the density of the absorbing material. It will be seen from *Figure 5.6* that

$$\varrho x = \frac{\text{Mass}}{\text{Volume}} \times x$$

$$= \frac{\text{Mass}}{\text{Area}} = m_a$$

therefore $\qquad I_x = I_0 e^{-\mu_m m_a}$

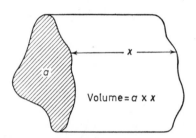

Figure 5.6. *Concept of equivalent thickness m_a as mass per unit area of absorber:*

$$x = \frac{volume}{Area\ a}$$

$$m_a = \varrho x = \frac{Mass}{Volume} \cdot \frac{Volume}{Area} = \frac{Mass}{Area}$$

At first the above may seem a rather cumbersome elaboration, but in practical radiology it is very convenient, and the mass absorption coefficients of the elements, for various radiation wavelengths, are given in tables of physical and chemical constants.

5.2.2. *The Principles of X-ray Filters*

In any material, x-ray absorption always occurs by Compton scattering and photo-electric energy transfer, but this absorption decreases as the energy of the radiation photons increases. Photons with sufficient energy to eject a *K* shell electron from atoms of the absorbing material will, in addition, be capable of absorption by this process.

If a graph is made of the linear absorption coefficient μ for a particular absorber against x-ray photon energy, the general trend will be, of course, a decrease (the greater the photon energy the greater the penetration). However there is a sharp increase of

absorption at the ejection energy for a *K* shell electron, E_K in
Figure 5.7. This steep increase is referred to as the *K-edge* for the
element concerned. The *K*-edges of different elements occur at
different, characteristic, energies and this property enables materials
to be used as filters for x-radiation. It can be inferred from the
above discussion about *K*-edges that there will also be *L*- and *M*-
edges but these need not be further considered in this discussion
of principles.

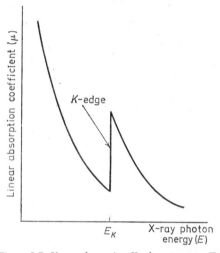

Figure 5.7. *X-ray absorption K-edge at energy E_K*

Filter materials would be chosen, one having its *K*-edge of
slightly higher energy than that of the *K* spectral line of the target
element being used—(*K*-edge)$_1$ in *Figure 5.8*—and another having
its *K*-edge of somewhat lower energy—(*K*-edge)$_2$ in *Figure 5.8*.
The result would be that much of the background white radiation,
and other spectral lines, would be absorbed.

As an example of the use of filters, suppose it is required to
provide monochromatic x-radiation within the wave band 1×10^{-10}
to 2×10^{-10} m. The K_α line of copper is known to have wavelength
1.54×10^{-10} m (photon energy 8.05 keV or about 1.30×10^{-15} J)
and is suitable. The *K*-edge of nickel corresponds to wavelength
1.49×10^{-10} m (photon energy 8.33 keV or about 1.33×10^{-15} J)
and that of iron to wavelength 1.74×10^{-10} m (photon energy
7.11 keV or about 1.14×10^{-15} J). Thus nickel and iron filters of

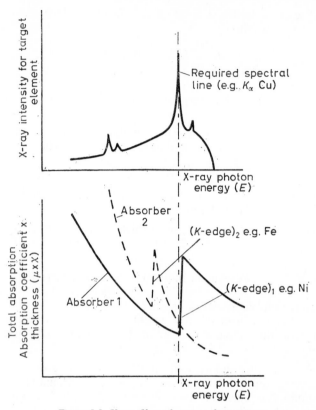

Figure 5.8. X-ray filters for monochromatiser

appropriate thicknesses (each, of course, has a different μ value) will suit the requirement for monochromatism, and the above figures correspond approximately to the situation illustrated in *Figure 5.8*.

5.3. QUESTIONS

1. The potential difference across an x-ray tube is 12.5 kV. What is the shortest wavelength emitted by the tube?

2. In the x-ray tube of Question 1 what is the de Broglie wavelength of the electron beam?

95

3. The characteristic K lines of the silicon x-ray spectrum have wavelengths as follows:

$$K_\alpha \quad 7.125\ 4\times10^{-10} \text{ m}$$
$$K_\beta \quad 6.768\ 1\times10^{-10} \text{ m}$$

What are the photon energies of the radiations making up these lines?

4. What is the minimum potential difference across an x-ray tube that will produce the K line spectrum for a silicon target?

5. For monochromatic x-rays of wavelength 2×10^{-11} m the mass absorption coefficient of lead is 0.490 m² kg⁻¹. What is the half-thickness of lead for this particular radiation wavelength? (The density of lead is 1.130×10^4 kg m⁻³.)

6. Referring to Question 5, what thickness of lead shielding will reduce the intensity of the 2×10^{-11} m wavelength radiation by 98%?

7. For 2.00 MeV gamma radiation, the mass absorption coefficient of iron is 4.24×10^{-3} m² kg⁻¹. (*a*) What is the half-thickness of iron for this particular radiation?

(*b*) What thickness of iron would be required to reduce the intensity of this radiation to 1/32 of its incident value? (Take the density of iron to be 7.80×10^3 kg m⁻³.)

8. An industrial radiographic unit is weld-testing on a bridge construction project using a source of monochromatic gamma radiation. The operator uses shielding 1.5×10^{-2} m thick of material whose linear absorption coefficient is 1.80×10^2 m⁻¹. What percentage reduction in intensity of this particular radiation does the shielding provide?

Chapter 6

Groups and Ordered Assemblies of Atoms

6.1. INTER-ATOMIC BONDS

6.1.1. General

In bulk materials in the solid state the component atoms are obviously more or less strongly bonded together to form the whole. It is these bonds which are discussed here. Inter-atomic bonds are classified into four main types, but although there are fairly clear cut examples of each type, it should be borne in mind that in most real materials the naturally occurring bonds show evidence of more than one type.

The thermal energy possessed by any solid material is represented by the vibrational kinetic energy of its constituent atoms, ions or molecules, which oscillate about a mean position, as well as the kinetic energy of any unattached electrons which may exist in the material. The closeness of the packing of atoms, ions or molecules in the solid state therefore tends to be restricted by this thermal vibration. The removal of heat reduces the vibrational energy and encourages closer packing. Conversely if heat is added to a solid the packing becomes less close and expansion is generally observed. If sufficient heat is put into any inorganic solid the vibrational energy will become so great that it exceeds the bonding energy, the atoms, ions or molecules 'fall apart' and the solid is said to melt.

6.1.2. The Ionic Bond

In this bond the component atoms, after assuming the ionic state, are held in combination by electrostatic coulomb forces in the following manner.

It will be recalled (Section 3.4.) that the valence electron octet (or *K* shell duplet for atoms near He in the list of elements) represents a highly stable atomic state. Reference to *Table 3.1.* shows that such elements as sodium (Na), potassium (K), rubidium (Rb) etc. (Group IA of the periodic table) will attain the valence octet structure by losing one valence electron thus becoming ionised as Na^+, K^+ and Rb^+ etc. On the other hand elements such as fluorine (F) and chlorine (Cl), (Group VIIA) will attain the octet structure by gaining one valence electron and becoming ionised as F^- and Cl^-.

If, for example, atoms of sodium and of chlorine are brought

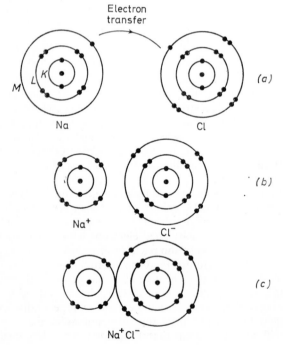

Figure 6.1. Schematic representation of formation of ionic bond in NaCl

98

together there will occur a valence electron transfer from the sodium to the chlorine thereby producing the two ions Na^+ and Cl^- each with the low energy valence electron octet. These ions will be mutually attracted, then held together, by electrostatic coulomb forces, as in *Figure 6.1*. These powerful forces constitute the *ionic bond* and in the above case give rise to the compound material sodium chloride (NaCl, or more correctly Na^+Cl^- since the constituents exist as ions in the solid material). The great strength of this bond is suggested by the relatively high melting-point temperatures of ionic solids which indicates the large amount of energy (as heat) necessary to disrupt the bond (*see* Sub-section 6.1.1).

Ionic bonds also occur with divalent elements, for example in magnesium chloride, two valence electrons transfer from the magnesium atom, one going to each of two chlorine atoms and the resulting ions Mg^{++}, Cl^- and Cl^- are bonded by electrostatic forces. In the case of the ferrous oxide FeO two electrons transfer from the iron atom to the oxygen atom, but such bonds are not purely ionic.

The constituents of ionic materials tend always to remain as ions, and if an ionic solid is dissolved in a suitable liquid, or if it is fused (melted), the ions acquire the ability to move freely and the liquid will be a good electrical conductor by virtue of this ionic mobility. Such conducting liquids are called *electrolytes*.

6.1.3. The Covalent Bond

In the covalent bond there is a sharing of valence electrons between two, or more, atoms in such a manner that each participating atom effectively achieves the valence octet (or duplet in the case of hydrogen) associated with high stability. When such electron sharing takes place it must occur in accordance with the Pauli exclusion principle which has a particular significance with respect to electron spin, as indicated in the case of hydrogen dealt with below (*see also* Sub-section 3.3.6).

When it is not possible to form complete electron octets by the sharing of one pair of electrons it can happen that two pairs or even three pairs of electrons are shared, giving *double* and *triple* bonds respectively.

A simple case of the covalent bond is that between two hydrogen atoms. The sharing of the two electrons effectively provides each atom with the *K* shell duplet associated with the low energy helium structure. To meet the requirement of the Pauli exclusion principle it is necessary that the two electrons have opposite spin, and this limits the choice of partner for a given atom to one having an electron with opposite spin.

Figure 6.2 illustrates several examples of the covalent bond from which it will be inferred that such elements as hydrogen, fluorine,

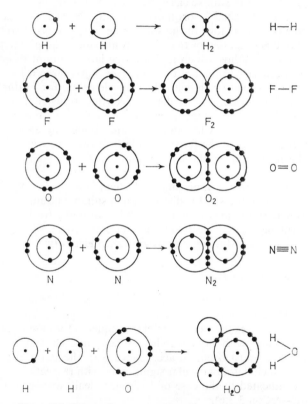

Figure 6.2. Schematic representation of various covalent bonds

oxygen, nitrogen, etc. are unlikely to exist freely in the atomic state purely from energy considerations. They are, in fact, found as diatomic units. Water, which is compounded of two elements exists as a triatomic unit of two hydrogen and one oxygen atoms. Such diatomic and triatomic units are typical *molecules*, of which more will be said later.

As an illustration of how a natural bond may exhibit characteristics of more than one of the basic types, the bond between two hydrogen atoms, although predominantly covalent, provides evidence of the ionic type. At any instant there is a finite probability that the two electrons of the molecule will both be more closely associated with one of the hydrogen nuclei than with the other. Thus instantaneously there is a negative ion (H^-) and a positive ion (H^+). These two ions will obviously be attracted electrostatically in the manner of the ionic bond. This situation is illustrated in *Figure 6.3* which is purely schematic.

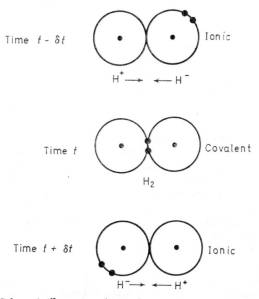

Figure 6.3. Schematic illustration of ionic characteristic in predominantly covalent bond (hydrogen molecule)

6.1.4. The Metallic Bond

About two thirds of all the elements are metals and they can be considered as falling into three main groups.

(*a*) Alkali metals such as lithium, sodium and potassium and alkaline earth metals such as beryllium, magnesium and calcium, all of which have outer *s* electrons.

(*b*) Transition metals such as titanium, vanadium and chromium all of which contain *d* or *f* electrons in incomplete inner shells, and which exhibit several valencies. Another example is iron which has six *d* electrons in its *M* shell, the maximum complement being ten. It exhibits valencies 2 and 3 because the energy step is small for an *M* shell *d* electron to transfer to the *N* shell *p* state (the *N* shell *s* state already having its full complement). Reference to *Table 3.1* and Section 3.4 will help make this clear.

(*c*) Amphoteric and 'semi'-metals which are those having outer *p* electrons, for example gallium, germanium, arsenic, indium, tin, antimony, thallium, lead and bismuth. These possess some metallic properties and their metallic nature (the degree of metallic type bonding) increases with a decrease in the number of outer *p* electrons. Also, for a given number of outer *p* electrons, the metallic nature increases with increasing atomic number. For example, tin is more metallic than antimony, but antimony is more metallic than arsenic. Reference to *Table 3.1* is essential in following this.

The metallic type of bond, unlike the ionic and covalent bonds, cannot exist between two, three or four atoms but involves a closely packed assembly of a large number of atoms. Pure, solid sodium provides an example of the simplest form of the metallic bond. In the solid state sodium atoms achieve the low energy valence octet by each liberating its single 3*s* valence electron. These freed electrons remain amongst the assembly which now comprises Na^+ ions, without being in any way tied to a particular ion. They therefore wander freely amongst the ions with a random motion whose kinetic energy represents a major part of the thermal energy of the metal.

In the solid state the packing of the metallic ions is sufficiently close for these freely wandering electrons to be, on average, slightly closer to the nearest nucleus than the valence 3*s* electron is in an isolated atom. Recalling that the energy of an atom is reduced when an orbiting electron moves towards the nucleus (*see* Sub-section 3.3.3) it is obvious that the solid state, as described, represents a lower energy condition than either a liquid or vapour which each comprise assemblies of relatively more isolated atoms.

The freedom of movement of the liberated electrons amongst the assembly of metal ions is very considerable. This is not difficult to visualise considering that an electron is around 10^{-4} times the size of a metal ion. Thus there exists the very useful model of a solid metal comprising an ordered, three-dimensional array of closely packed metallic ions, randomly distributed amongst which are unattached electrons more or less free to wander. Such a model gives rise to the electron gas theory in which all the free electrons can be treated in terms of the gas laws for some purposes.

As a generalisation, the valence electrons of the transition metals are around 30% more tightly bound than those of the alkali and alkaline earth metals. For this reason, amongst others, the solid state bond in the transition metals is not so straightforward as the ideal metallic bond described above. In the solid state the transition elements generally have their atoms sufficiently closely packed for incomplete energy levels of neighbouring atoms, such as $3d$ and $4s$, to overlap, thereby encouraging some covalent bonding to complete the electron complement of these energy levels. This tends to reduce the number of valence electrons released as free electrons, and hence the degree of metallic bonding. This also explains why the number of free electrons in most metals, when expressed as the number contributed per atom, is not a whole number. Most of the transition elements thus exhibit considerably lower electrical and thermal conductivity than would be expected if all valence electrons were released as free conduction electrons.

The third group of metals, the amphoteric or semi-metals, having outer p electrons, bond such that those with one outer p electron show typical metallic characteristics, as for example, aluminium. But with increasing numbers of outer p electrons there is an increasing tendency towards the covalent bond and consequently a reduction in metallic characteristics.

6.1.5. van der Waals' Bond

van der Waals' bond is very weak compared with the ionic, covalent and metallic bonds. It is attributable to the fact that at any instant, in any atom or atom group there will be a charge separation (i.e. a dipole effect). In other words the centre of positive charge will not exactly coincide with the centre of negative charge and the atom or group will behave as a rapidly fluctuating dipole. In the time-averaged sense this dipole effect is zero in all atoms and in many atom groups and these are termed *non-polar*. Nevertheless the fluctuating dipole effect can give rise to mutual electrostatic forces between single atoms and between non-polar groups. These weak forces are attractive and constitute the so-called van der Waals' bond.

103

In other, particularly non-symmetrical, atom groups, in addition to the rapidly fluctuating dipole effect just described there is often a larger scale permanent (non-fluctuating) charge separation giving a permanent dipole. Such groups are called *polar*; and in communities of polar groups, in addition to the van der Waals' forces, there are other much stronger electrostatic forces to be taken into account.

6.2. MOLECULES

6.2.1. General

By definition a molecule is the smallest unit of a material capable of existing independently whilst retaining the identity of that material. Molecules are most commonly small groups of atoms, although it will be realised that certain atoms are themselves also molecules, for example the atoms of the inert gases helium, neon, argon, krypton, xenon and radon. Such single atom molecules are termed *monatomic*.

Molecules of certain other materials have two atoms, either two of the same element such as hydrogen (H_2) and oxygen (O_2), or of two different elements forming chemical compounds such as carbon monoxide (CO) and hydrogen chloride (HCl); two atom molecules are termed *diatomic*. Molecules of yet other materials are *triatomic*, such as water (H_2O), and others again have many atoms and are termed *polyatomic*. Some of these polyatomic molecules have very large numbers of atoms indeed.

6.2.2. Polar Molecules and Directional Bonds

In Sub-section 6.1.5 mention was made of polar molecules, meaning those exhibiting a permanent measurable dipole due to non-uniform charge distribution.

The measure of the dipole strength is the *dipole moment* which represents the product of the charge and the distance of charge separation. The units of dipole moment are coulomb metres (C m).

Those molecules with the ionic bond, and many covalent bonded molecules, exhibit permanent dipole moments. An example of the former is the hydrogen chloride molecule HCl, and of the latter the water molecule H_2O. *Table 3.1* gives some examples of molecular dipole moments as measured in the gaseous state.

Table 6.1 MOLECULAR DIPOLE MOMENTS (Gaseous State)

Molecule		Dipole moment
Hydrogen chloride	HCl	3.6×10^{-30} C m
Water	H_2O	6.2×10^{-30} C m
Potassium fluoride	KF	24.4×10^{-30} C m
Formaldehyde	CH_2O	7.3×10^{-30} C m

It is not difficult to form a mental picture and to understand the charge separation in the case of ionic molecules, as will be seen from *Figure 6.4* which represents the hydrogen chloride molecule H^+Cl^-.

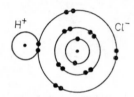

Figure 6.4. Dipole molecule (HCl)

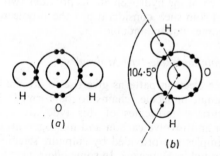

Figure 6.5. The water molecule: (a) poor representation of H_2O molecule: (b) good representation of H_2O molecule indicating directional bonds

In many covalent molecules, however, the charge separation occurs because of the specific geometry of the molecule. For example, the water molecule is not as represented in *Figure 6.5(a)* but is rather as in *Figure 6.5(b)*, which illustrates the specific directional nature of the covalent bond.

Again in the case of the formaldehyde molecule CH_2O the bonding is not random but is specifically directional. The molecular form of formaldehyde is always as illustrated in *Figure 6.6*. The

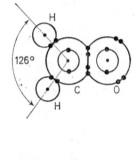

Figure 6.6. *The formaldehyde molecule, CH_2O*

carbon atom shares two pairs of electrons with the oxygen atom giving a double bond (*see* Sub-section 6.1.3), and one electron shared with each of the hydrogen atoms provides two single bonds as illustrated. From such considerations the dipole nature of the molecules becomes understandable.

6.2.3. Long-chain and Ring-type Molecules

Certain valence electron patterns give rise to atomic bonds which allow long, single-molecule, chains to be formed. Most of the naturally occurring molecules of this type are compounds of organic origin and involve carbon and hydrogen atoms, however, elemental examples are provided by sulphur, selenium and tellurium, as illustrated in *Figure 6.7*. In some *allotropic** forms sulphur

---

* Allotropes—different physical forms of the same chemical substance.

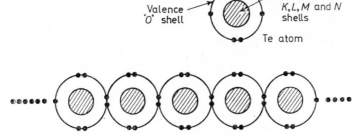

Figure 6.7. *The tellurium molecule. (Sulphur and selenium can adopt a similar formation)*

exists as closed ring molecules (S_8), as illustrated schematically in *Figure 6.8*, and a similar molecular ring structure is sometimes adopted by selenium. This type of bonding, when it does not result in a ring formation, may lead to open-ended molecules and these can be very long indeed.

Examples of long-chain organic groups consisting of carbon and hydrogen (hydrocarbons) are butane (C_4H_{10}) and octane (C_8H_{18}), illustrated in *Figure 6.9*, and polyethylene (C_nH_{2n+2} where n is a very large integer—say 10^4 to 10^6) is another example which is very long indeed. Benzene is a ring form of hydrocarbon

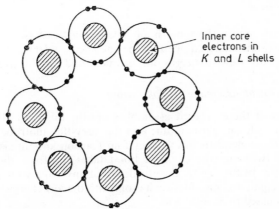

Figure 6.8. *Ring molecule (S_8) of sulphur. (This form can also be adopted by selenium)*

107

involving double bonds of the type to be discussed in Sub-section 6.3.2, and is illustrated in *Figure 6.10*. Very complex molecular structures can result from the substitution of other atoms or atom groups for one or more of the hydrogen atoms in such long-chain or ring-type hydrocarbon molecules.

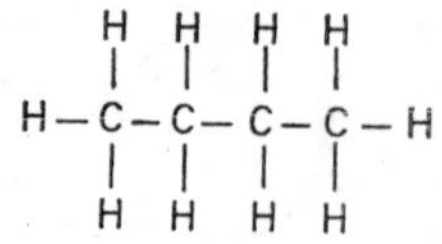

Figure 6.9. Structural formula of butane (C_4H_{10})

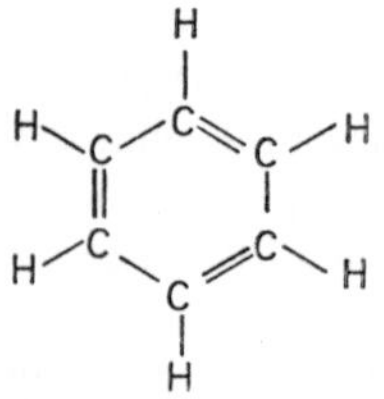

Figure 6.10. Structural formula of benzene ring (C_6H_6)

6.3. ORDERED ASSEMBLIES OF LIKE ATOMS

6.3.1. General

By its very nature the ionic bond involves at least two types of atom, so ionic materials cannot be included under the heading of 'like atoms' which concerns only covalent and metallic materials.

6.3.2. Covalent Assemblies of Like Atoms

Examples of these ordered assemblies are the diamond allotrope of carbon; silicon; germanium; and the 'grey-tin' allotrope of tin. Each of these elements has valency four (carbon and tin also have valencies two) and the covalent bond takes the form shown in *Figure 6.11*. In this figure, the covalency of any atom with its four nearest neighbours provides each atom with the low energy valence octet discussed in Section 3.4.

Because each atom needs four close neighbours it is said to have a *coordination number* 4. In three dimensions (the normal solid

108

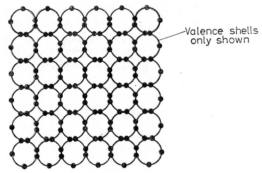

Valence shells only shown

Figure 6.11. Two-dimensional representation of covalent bonding in diamond, silicon, germanium and 'grey' tin

state condition) each atom has its four closest neighbours orientated in a tetrahedral formation as in *Figure 6.12*, and the tetrahedron is referred to as the *coordination polyhedron* of this type of structure.

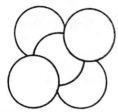

Figure 6.12. Tetrahedral arrangement of atoms for three-dimensional covalent bonding—coordination number 4

In germanium (Ge) and silicon (Si) the covalent bond strength is not particularly great. Normal ambient temperatures represent enough heat energy to break a few of the bonds, thereby releasing some electrons which can then contribute to electrical and thermal conduction. Such electrical conduction is termed *semi-conduction* because it is not of a high order; it is also termed *intrinsic* because it is a natural attribute of the material. Intrinsic semi-conduction will obviously increase as more bonds are broken by increasing the heat energy, so intrinsic semi-conductors exhibit a negative temperature coefficient of resistance. Their conductivity at the absolute zero of temperature is theoretically zero.

109

In graphite—which is another allotropic form of carbon—the atoms are known to form into planes of hexagonal 'ring' structure. The appropriate number of bonds can be achieved by double bonds occurring in the rings, as indicated in *Figure 6.13*. These double bonds are not confined to particular positions, such as those

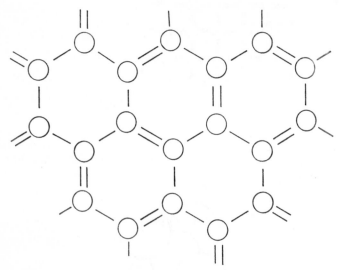

Figure 6.13. Schematic arrangement of hexagonal ring structure of graphite

shown, but frequently change position, or 'resonate', within the plane structure: at any instant, an atom will have one double and two single bonds. Expressed another way, for one third of its existence the bond between any two atoms will be a double bond. In this structure, because each atom has three nearest neighbours, the coordination number is 3 and the coordination polyhedron is in fact a plane triangle.

In this allotropic form of carbon the planes of hexagonal ring are held, one to another, only by weak van der Waals' forces as represented in *Figure 6.14*. These planes will very easily slip over one another and this accounts for the useful lubricating property of this form of carbon (and of other materials having a similar laminar structure such as molybdenum disulphide MoS_2).

Bearing in mind the resonant nature of the double bonds in the graphite structure and the consequent electron mobility, it is not difficult to visualise the comparative ease with which, by a 'switching' mechanism, an effective electron drift can occur in the plane of the hexagonal rings. The electrical and thermal conductivity of graphite is therefore good in the hexagonal planes, but there is negligible conductivity normal to these planes because of the nature of the van der Waals' bond.

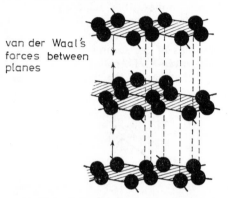

van der Waal's
forces between
planes

Figure 6.14. Structure of graphite, showing planes of hexagonal 'rings'. The planes are held together by van der Waals' forces

In a piece of common graphite, having its hexagonal planes quite randomly orientated, the thermal and electrical conductivity will be uniformly quite good. On the other hand, if the material is so worked that the hexagonal planes all orientate in a particular direction there will be marked thermal and electrical *anisotropy**.

6.3.3. Metallic Assemblies of Like Atoms

Since only assemblies of like atoms are to be considered, this section must refer solely to pure metals. It will be recalled that the metallic bond requires many atoms. It cannot exist between two or three or even a score or so metal atoms, and the concept of a metal comprising closely packed ions (atoms which have contributed electrons to the free electron 'gas') has already been established.

In the case of ordered, covalent, assemblies of like atoms such as solid germanium and the diamond allotrope of carbon, the coordination number is four, which means that each atom needs

* Anisotropy—different characteristics in different directions within a material.

111

four close neighbour atoms to fulfill the bond requirement. This need is satisfied in these materials by tetrahedral coordination sub-structures. In the metallic bond there is no specific requirement of the method of packing of the metal atoms (ions more strictly), other than that they pack as closely as they can.

Very close packing involves a coordination number 12, for which the polyhedra are complex. It is achieved by the layers of ions (considered as uniform spheres) stacking either *ABABA*... or *ABCABCABC*.... In the former, the first layer of the stack is called *A* and the ions of the second layer, *B*, rest in the natural depressions of the layer *A*. Then in the third layer the ions each lie directly above the ions of the first layer *A*. Such a stacking arrangement is commonly referred to as *close packed hexagonal* (CPH) and is illustrated in *Figure 6.15*.

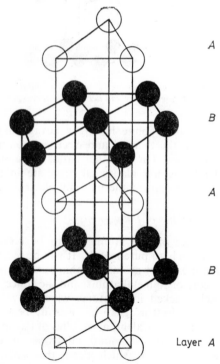

Figure 6.15. CPH stacking sequence ABABA.... (coordination number 12)
('Exploded' view—atoms to be considered nearly touching)

In the case of *ABCABC*.... stacking, the layers *A* and *B* stack as before but the third layer ions rest in depressions of the layer *B* which are *not* directly above ions of layer *A*. This type of close packing is referred to as *Face Centred Cubic* (FCC) because a unit cube, orientated at 45° to the layers *ABCABCA*.... exists with an ion at the centre of each face. This stacking is illustrated in *Figure 6.16*.

There is another type of metallic ion stacking which is not quite so close-packed as the two just dealt with, this is the *Body Centred Cubic* type (BCC). In this arrangement each atom has eight closest

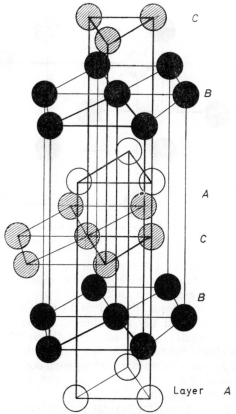

Figure 6.16. FCC stacking sequence ABCABCA.... *(coordination number 12)*
('Exploded' view—atoms to be considered nearly touching)

113

neighbours and therefore a coordination number 8, and the coordination polyhedron is a cube. BCC stacking is illustrated in *Figure 6.17*. The factors put forward as causing the adoption of this formation rather than the more closely packed CPH and FCC formations (which might be expected to provide a lower energy

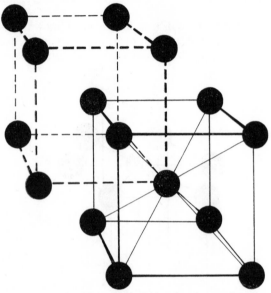

Figure 6.17. The BCC structure (coordination number 8)
('Exploded' view—atoms to be considered nearly touching)

system—*see* Sub-section 6.1.4) are (*i*) that the vibrations of thermal energy prevent the closest possible packing (deduced from the fact that many metals which adopt BCC packing at normal ambient temperatures will transform to CPH or FCC packing at much lower temperatures), and (*ii*) the influence of partial covalent bonding in many metals (*see* Sub-section 6.1.4) may tend to reduce the number of closest neighbours needed. Also there may be some effect due to the directional nature of this partial covalent bonding. The fact that most transition metals solidify with a BCC structure supports these latter suggestions.

The packing formations for several pure metals are given in *Table 6.2* which is mainly restricted to the more common industrial metals at normal ambient temperatures.

114

Table 6.2 PACKING FORMATION OF SOME PURE METALS

Atomic no.	Metal	Packing	Atomic no.	Metal	Packing
11	Sodium	BCC	28	Nickel	FCC
12	Magnesium	CPH	29	Copper	FCC
13	Aluminium	FCC	30	Zinc	CPH
19	Potassium	BCC	47	Silver	FCC
20	Calcium	FCC	48	Cadmium	CPH
21	Scandium	CPH	74	Tungsten	BCC
22	Titanium	CPH	77	Iridium	FCC
24	Chromium	BCC	78	Platinum	FCC
26	Iron	BCC	79	Gold	FCC
27	Cobalt	CPH	82	Lead	FCC

6.4. ORDERED ASSEMBLIES OF UNLIKE ATOMS

6.4.1. General

In the case of metallic assemblies of like atoms, all atoms being the same size, the close atomic packing involves one of the three regular patterns CPH, FCC or BCC. Covalent assemblies of like atoms involve directional bonds and these largely determine the atomic arrangements, although uniformity of atomic size also contributes to their structural pattern.

In the case of unequal sized atoms or ions the packing arrangements can vary much more widely although classification into a relatively few basic geometrical units, depending upon atom or ion sizes, can be achieved.

6.4.2. Ordered Ionic Assemblies

The ionic bond is in no way directional, so the nature of the bond itself places little or no restriction on the atomic packing arrangements, which are therefore almost entirely dependent upon the relative size factor.

The number of anions (negative ions—*see* Sub-section 3.1.5) closely associated with each cation (positive ion) is called the *ligancy* of an ionic system. The ligancy is thus the number of closest neighbour anions of one element surrounding each smaller cation of the other element, and is analogous to coordination number in covalent and metallic structures.

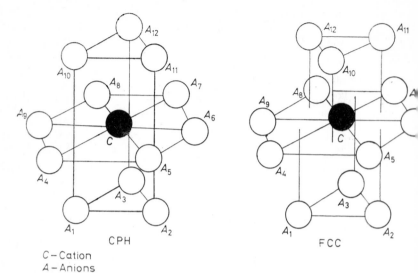

C — Cation
A — Anions

Figure 6.18. Cation/anion ratio 1.0, representing hypothetical ligancy 12 (actual coordination number 12)

('Exploded' view—ions to be considered nearly touching)

C-Cations
A-Anions

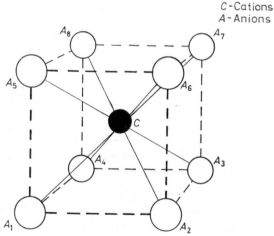

Figure 6.19. Cation/anion ratio about 0.75, ligancy 8. Cubic ligancy polyhedron

('Exploded' view—ions to be considered nearly touching)

116

The relative ion sizes are usually expressed as the ratio
$\frac{\text{Radius of cation}}{\text{Radius of anion}}$. If the two radii were to be the same the ratio
would be 1.0, the ligancy would be twelve and the packing would
be either CPH or FCC as in *Figure 6.18* (*see also* Sub-section 6.3.2).

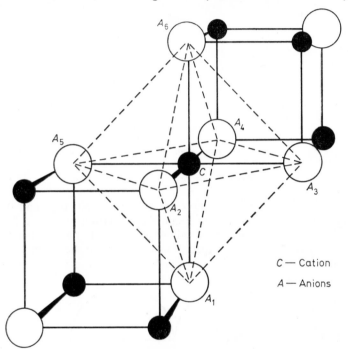

*Figure 6.20. Cation/anion ratio about 0.5, ligancy 6. Octahedral ligancy poly-
hedron*

('Exploded' view—ions to be considered nearly touching)

This type of packing, in fact, does not occur in ionic materials
because the anion and cation radii are never equal. For ionic
materials the radius ratio will, practically without exception, be
less than 1.0.

From purely geometrical considerations, for ratios between
1.0 and about 0.73 the ligancy is expected, by calculation, to be 8;
the anion packing would then be *cubic* around the cation as in
Figure 6.19 (analogous to the BCC structure described in Sub-

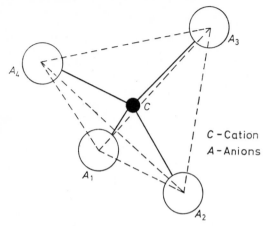

Figure 6.21. Cation/anion ratio about 0.38, ligancy 4. Tetrahedral ligancy polyhedron

('Exploded' view—ions to be considered nearly touching)

section 6.3.2). When the ratio lies between about 0.73 and 0.41 the ligancy is expected to be 6 and the formation of anions about a cation *octahedral*, as illustrated in *Figure 6.20*. For a ratio between about 0.41 and 0.23 the expected ligancy is 4 and the anion formation is *tetrahedral*, as illustrated in *Figure 6.21*. For ratios between

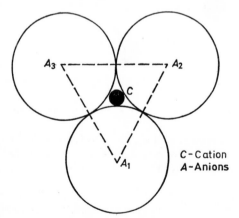

Figure 6.22. Cation/anion ratio about 0.16, ligancy 3. Planar trigonal (triangular) ligancy 'polyhedron'

118

0.23 and 0.16 the expected ligancy is 3 and the anion formation *planar trigonal*, i.e. *triangular*—the anion units enclosing each cation being in two dimensions as in *Figure 6.22*. Finally for a ratio between 0.16 and 0.00 the formation is expected to be *linear* associated with ligancy 2, as in *Figure 6.23*. The stereometry* of the anions in relation to each cation provides *ligancy polyhedra* which are analogous to coordination polyhedra (*see* Sub-section 6.3.2).

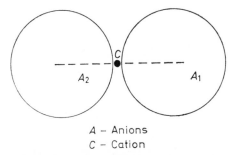

A – Anions
C – Cation

Figure 6.23. Cation/anion ratio about 0.07, ligancy 2. Linear ligancy 'polyhedron'

Some examples of ionic compounds, indicating the radius ratio, the expected ligancy and the *observed* structure of the ligancy polyhedra, are given in *Table 6.3*.

Table 6.3 SOME IONIC COMPOUNDS

Compound	Ratio $\dfrac{Cation\ radius}{Anion\ radius}$	Expected ligancy	Observed structure of ligancy polyhedron
BeO	0.23	4	Tetrahedral
MgO	0.47	6	Octahedral
CsCl	0.93	8	Cubic

6.4.3. Ordered Covalent Assemblies of Unlike Atoms

In terms of utilisation, probably the most important assemblies of this type are those which involve long-chain molecules. Those long-chain molecules which are very regular in their own structure can most easily form ordered assemblies in the solid state. Examples of these are polyethylene, polyvinylidene chloride and polytetra-fluoroethylene, *Figure 6.24*. The first of these has as its basic unit

* Stereometry—solid, or three-dimensional, geometry.

119

the *mer* CH_2 which has the structural formula $-\overset{\displaystyle H}{\underset{\displaystyle H}{C}}-$, so that the

molecule is a 'multi-mer', more commonly called a *polymer*.

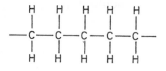

(a) Polyethylene

(b) Polyvinylidene chloride

(c) Polytetrafluoroethylene

Figure 6.24. Regular polymer examples

In the second of the above long-chain molecules there is regularly ordered substitution of chlorine atoms (Cl) for half of the number of hydrogen atoms of the simpler polymer polyethylene, and the mer is $-\overset{\displaystyle H}{\underset{\displaystyle H}{C}}-\overset{\displaystyle Cl}{\underset{\displaystyle Cl}{C}}-$. In the last named there is complete substitution of fluorine (F) for the hydrogen of the polyethylene molecule, so the mer here is $-\overset{\displaystyle F}{\underset{\displaystyle F}{C}}-\overset{\displaystyle F}{\underset{\displaystyle F}{C}}-$.

Many of the long-chain hydrocarbon molecules, and especially those with substitution groups, are so long and involved that in the solid state they remain in a state of great disorder and entanglement. Thus their structure is only ordered in the short-range sense that the molecules themselves represent some degree of order. The subject of these materials more properly belongs in Chapter 8, where they are dealt with in some detail.

6.4.4. Ordered Metallic Assemblies of Unlike Atoms

Only one type of assemblage will be dealt with in this sub-section. It is restricted to two metallic elements whose individual, elemental, atomic packing patterns are of the same type (FCC, CPH or BCC, for example), and the atoms of which are well within 15% of being the same size (the relative size limit for substitutional alloying— *see* Sub-section 9.4.2).

In such a system, straightforward substitution of one atom or ion for another can occur. If the atoms or ions of the two component metallic elements are very close in size, not only will the structure of the assemblage be the same as that of the individual components but the distribution of one component amongst the atoms or ions of the other will be quite random. For example, elemental copper and nickel both have FCC structures and their ionic radii are within about 4% of each other. When alloyed in any proportions, the alloy structure is also FCC and the distribution of nickel amongst the copper (or of copper amongst the nickel) is completely random. *Figure 6.25* illustrates this.

If the individual packing patterns are of the same type but, on the other hand, the atoms of the two elements are near to the limit

Figure 6.25. Overall order but random distribution of two species

of size difference for substitutional alloying (about 15%), the overall structural pattern will still be that of the individual elements (e.g. FCC, CPH or BCC), but random distribution of atoms of one element amongst those of the other will cause maximum distortion of the structure. To minimise this distortion (thereby minimising the energy of the system) a second degree of order will often occur, over and above the general structural pattern. Each component then presents an ordered pattern in its own right and *super order* is said to exist. A structure possessing such super order is termed a *super-lattice* and is represented in *Figure 6.26*.

Figure 6.26. Overall order plus ordering with respect to a species: a super-lattice

6.4.5. Ordered Covalent–Ionic Assemblies

Oxygen and silicon are the two most abundant elements of Earth's composition and silicon-oxygen compounds are therefore good examples to take of this type, being extremely common. The compound silica (SiO_2) exists naturally in many allotropic forms, for example as the pure forms quartz, cristobalite and tridymite, and in combination with water as the amorphous minerals opal and flint. This compound has an ionic structure although its bonding is predominantly covalent. The cation to anion radius ratio of silica is $0.42/1.32 = 0.31$, and reference to Sub-section 6.4.2 indicates an expected ligancy of 4, hence a tetrahedral ligancy polyhedron, which in fact it does possess.

These tetrahedra comprise $(SiO_4)^{4-}$ units, each consisting of a central Si^{4+} ion with tetrahedrally disposed O^{2-} ions, and join one with another in such a manner that each O^{2-} ion is shared between two tetrahedra, leading to the correct overall chemical composition SiO_2 (i.e. one Si^{4+} to two O^{2-} ions).

122

Geometrically the tetrahedra usually join corner to corner with a fairly open (i.e. not closely packed) structure, and the various possibilities of this lead to the different allotropic forms of the compound. Often these allotropes possess long-range order and are crystalline.

Silicates are those materials which involve various metallic elements in combination with such silicon-oxygen anionic groups as $(SiO_3)^{2-}$, $(Si_2O_5)^{2-}$ and $(Si_4O_{11})^{6-}$. Examples of silicates are the minerals enstatite $Mg(SiO_3)$, beryl $Be_3Al_2(SiO_3)_6$, kaolinite $Al_2(OH)_4(Si_2O_5)$—a *clay*, talc $Mg_3(OH)_2(Si_2O_5)_2$ and tremolite $Ca_2Mg_5(OH)_2(Si_4O_{11})$. In the *mica* silicate K $Al_2(OH)_2(Al\ Si_3O_{10})$, one silicon ion in every four of the silicon-oxygen group $(Si_2O_5)^{2-}$ is replaced by an aluminium ion to produce the $(Al\ Si_3O_{10})^{5-}$ anionic group.

Kaolinite, talc and mica are typical examples of layer-type structure comprising very strongly bonded layers or sheets which are in turn weakly bonded to each other by Van der Waals' forces, rather like those of the graphite structure (*see* Sub-section 6.3.2). These materials can therefore be expected to have similar lubricating properties to graphite, and indeed powdered talc (commonly called 'french chalk') and mica are both suitable for dry lubrication, and wet clay is notoriously slippery.

6.5. SHORT-RANGE AND LONG-RANGE ORDER

The term 'short-range order' refers to ordered arrangements within the compass of a small number of atoms or molecules (as represented, for example, the by atomic order of individual large molecules, such as the long-chain molecules; or by the coordination or ligancy polyhedra discussed in Sections 6.3 and 6.4). 'Long-range order', on the other hand, implies an ordered arrangement involving very large numbers of atoms or molecules.

When the structural order in a material is long-range the material is said to be *crystalline* and its three-dimensional, geometrical,. structural matrix is called a *lattice*. Most materials, whether ionic covalent or metallic, are crystalline.

When the ordering in a material is short-range, the material is said to be *amorphous**. An example would be a tangled, highly disordered, assemblage of very long chain molecules. The common form of polyvinyl chloride (PVC) is such a disordered assemblage.

---

* Amorphous—lacking form or shape.

Again, any molten material is, almost by definition, in a state of great disorder, so that if, for example, molten silica is sufficiently rapidly transformed to the solid state, there will be no time for diffusion to occur to the extent of achieving long-range order in the solid state. The resulting disordered solid is termed a *glass*. Flint, opal and obsidian are all natural glassy silicates.

6.6. QUESTIONS

1. If the mean electron drift velocity is 6.25×10^{-6} m s^{-1} along a copper rod of 10^{-4} m^2 cross section area, and the electric current is observed to be 10A, calculate a value for the density of free conduction electrons in copper.

2. Using the result of the calculation in Question 1 obtain a value for the average number of electrons contributed to the free electron gas by each atom of copper. Take the density of copper to be 8.92×10^3 kg m^{-3}.

Chapter 7

Crystalline Structure

7.1. INTRODUCTION

Atomic assemblages which possess long-range order are, by definition, crystalline. Crystalline structures are described in terms of
space lattices which are continuous three dimensional arrays of
points in space, having a repetitive geometrical pattern. In the most
widely used convention there are fourteen different types of space
lattice (A. Bravais, 1848), each of which represents a repetitive
pattern of a different basic unit. In fact there are only seven
fundamental lattice units (primitive cells), but several of these are
sub-classified to provide the fourteen Bravais lattices.

In crystalline structures the points in space of the lattices may
be occupied by atoms, ions or molecules.

7.2. THE BRAVAIS SPACE LATTICES

The seven fundamental lattice units are:

1. Cubic	4. Orthorhombic
2. Hexagonal	5. Rhombohedral
3. Tetragonal	6. Monoclinic

7. Triclinic

These are most easily described using three right-handed axes,
X, Y and Z, having angles alpha (α), beta (β) and gamma (γ)
between X and Y, Y and Z, and Z and X respectively. Unit lattice
spacings along these axes are represented by a along X, b along Y
and c along Z, as in *Figure 7.1*.

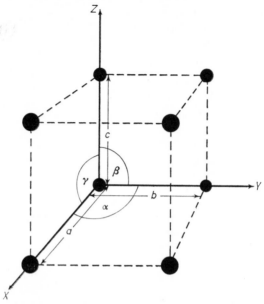

Figure 7.1. Crystallographic axes and lattice parameters

The sub-classifications of the fundamental lattice units are as follows:

Fundamental	*Sub-classification*
1. CUBIC	(*a*) Primitive
	(*b*) Body centred cubic (BCC)
	(*c*) Face centred cubic (FCC)
2. HEXAGONAL	(*a*) Primitive only (CPH)
3. TETRAGONAL	(*a*) Primitive
	(*b*) Body centred
4. ORTHORHOMBIC	(*a*) Primitive
	(*b*) Face centred—two opposite faces only
	(*c*) Body centred
	(*d*) Face centred—all faces
5. RHOMBOHEDRAL	(*a*) Primitive only
6. MONOCLINIC	(*a*) Primitive
	(*b*) Face centred—two opposite faces only
7. TRICLINIC	(*a*) Primitive only

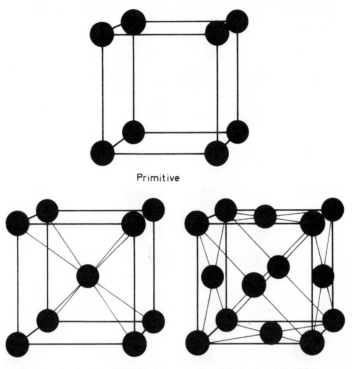

Primitive

Body centred cubic (BCC) Face centred cubic (FCC)

Figure 7.2. Fundamental cubic. $a = b = c, \alpha = \beta \neq \gamma$

The above are all illustrated in *Figures 7.2* to *7.8* and are the fourteen basic Bravais lattice types.

7.3. MILLERIAN CRYSTALLOGRAPHIC INDICES

When it is necessary to describe unambiguously the various planes and directions which exist in crystals, probably the most widely used convention is that originated and described by W. H. Miller in this *Treatise on Crystallography* (1839). This uses three- or four-digital indices for the purpose.

The Miller indices of a plane are obtained as follows. On the same cartesian axes used to describe the crystallographic space lattices, observe the distances along the axes X, Y and Z that the

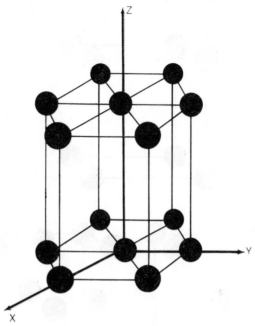

Figure 7.3. Fundamental hexagonal (primitive only) $a = b \neq c$, $\alpha = 120°$, $\beta = \gamma = 90°$

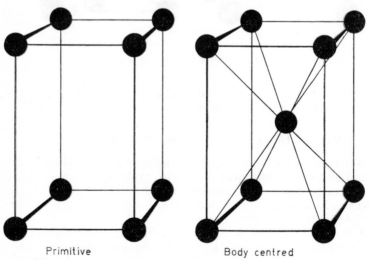

Primitive   Body centred

Figure 7.4. Fundamental tetragonal. $a = b \neq c$, $\alpha = \beta = \gamma = 90°$

128

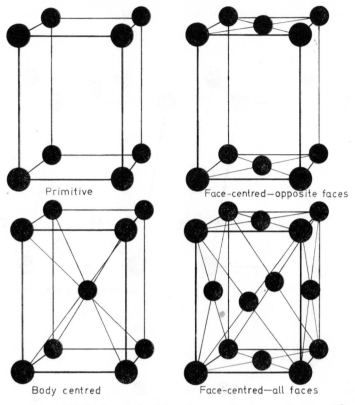

Figure 7.5. Fundamental orthorhombic. $a \neq b \neq c, \alpha = \beta = \gamma = 90°$

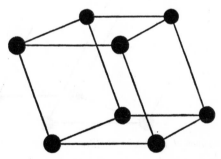

Figure 7.6. Fundamental rhombohedral (or trigonal)—primitive only. $a = b = c$,
$\alpha = \beta = \gamma \neq 90°$

129

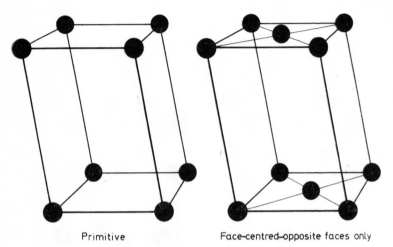

Primitive   Face-centred–opposite faces only

Figure 7.7. Fundamental monoclinic. $a \neq b \neq c, \alpha = \beta = 90° \neq \gamma$

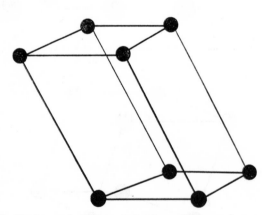

Figure 7.8. Fundamental triclinic (primitive only). $a \neq b \neq c, \ \alpha \neq \beta \neq \gamma$

130

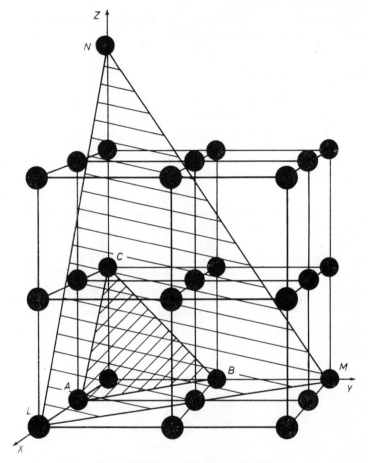

Figure 7.9. Planes (1 1 1) and (3 3 2)

required plane intersects the axes, measured in units of a, b and c. Take reciprocals of these distances, express them as the lowest integers which are in the same ratio, and enclose the resulting digits in parentheses.

For example, consider the plane A B C in *Figure 7.9*. The intersections with the axes X, Y and Z are respectively at $1 \times a$, $1 \times b$ and $1 \times c$, thus in units of a, b and c the intersections occur at 1, 1 and 1.

Taking reciprocals of these, expressing them as the lowest integers in the same ratio and placing them in parentheses gives

$$(1\ 1\ 1)$$

which are the Miller indices of the plane *A B C*, and of all planes parallel to it.

For another example consider the plane *L M N*. Intersections occur at $2a$, $2b$ and $3c$ so that the intersection distances in appropriate units are 2, 2 and 3. The reciprocals are $\frac{1}{2}$, $\frac{1}{2}$ and $\frac{1}{3}$ which, expressed as the lowest integers in the same ratio, are 3, 3 and 2. The Miller indices of this plane, and of all planes parallel to it, are thus

$$(3\ 3\ 2)$$

The generalised symbols for the three Miller indices of a plane are $(h\ k\ l)$, and some of the planes of a cubic system are illustrated in *Figure 7.10* together with their indices.

A four-number system is used to quickly identify hexagonal lattices, in which the fundamental Bravais unit has 8 faces compared with the 6 faces of all the other types (*see Figures 7.2 to 7.8*). The four numbers relate to axes *X*, *Y*, *W* and *Z*, as shown in *Figure 7.11*, and the symbols for the four indices are $(h\ k\ i\ l)$. There is a relationship between h, k and i such that,

$$h+k=-i$$

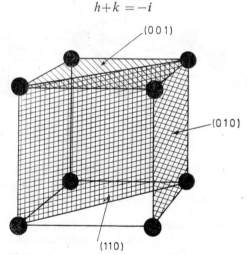

Figure 7.10. Further examples of Miller indices (h k l)

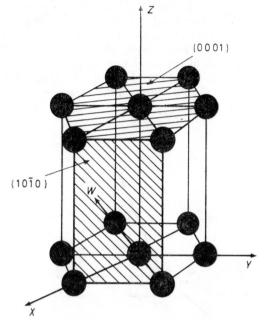

Figure 7.11. Miller indices (h k i l) for hexagonal system

and this will be readily understood by reference to *Figure 7.11*.
It will also be appreciated that the index *i* is not strictly necessary.

Consider the shaded basal plane in *Figure 7.11*: the intercepts with
the axes *X, Y* and *W* are all ∞ and with the *Z* axis the intercept is 1.
The values of *h, k, i* and *l* are the reciprocals of ∞, ∞, ∞ and 1
so the indices of this plane, and of all planes parallel to it, are

$$(0\ 0\ 0\ 1)$$

In a similar way the vertical shaded side of *Figure 7.11* has
indices*

$$(1\ 0\ \bar{1}\ 0)$$

since the *X, Y, W* and *Z* intercepts are respectively 1, ∞, −1 and ∞.

In a crystal, all planes of the same type are said to be of the same
family and a family of planes would be represented by the simplest

---

* Minus 1 is written $\bar{1}$ for convenience.

indices of the type enclosed in braces (i.e. 'curly' brackets). Thus in a cubic lattice the family $\{1\,0\,0\}$ consists of all the planes $(1\,0\,0)$, $(0\,1\,0)$, $(0\,0\,1)$, $(\bar{1}\,0\,0)$, $(0\,\bar{1}\,0)$ and $(0\,0\,\bar{1})$, which represent the six faces of the basic cube. Similarly the family $\{1\,\bar{1}\,0\,0\}$ refers to all planes of the same type in a hexagonal system and includes $(1\,\bar{1}\,0\,0)$, $(0\,1\,\bar{1}\,0)$, $(\bar{1}\,0\,1\,0)$, $(\bar{1}\,1\,0\,0)$, $(0\,\bar{1}\,1\,0)$ and $(1\,0\,\bar{1}\,0)$, the six similar sides of the primitive hexagonal unit.

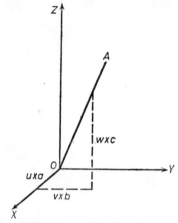

Figure 7.12. Directional indices [u v w] of OA

The indices of a direction in a crystal, such as *OA* in *Figure 7.12*, are obtained by taking the number of units of lengths a, b and c travelled successively from O in directions parallel to X, Y and Z to get from O to the line joining O and A. If these numbers are expressed as the smallest whole numbers in the same ratio, u, v and w, and are set in brackets ('square' brackets) thus

$$[u\,v\,w]$$

they are the Miller indices of the direction *OA*, and of all directions parallel to it. Indices giving the same 'type' of direction are said to be of the same *form* and comprise a set. To represent the set of a given form the simplest indices of the form are put in carats ('angle' brackets) thus the set

$$\langle 1\,1\,1 \rangle$$

comprises $[1\,1\,1]$, $[\bar{1}\,1\,1]$, $[1\,\bar{1}\,1]$, $[1\,1\,\bar{1}]$, $[\bar{1}\,\bar{1}\,1]$, $[1\,\bar{1}\,\bar{1}]$, $[\bar{1}\,1\,\bar{1}]$ and $[\bar{1}\,\bar{1}\,\bar{1}]$.

In all cubic systems a direction perpendicular to any plane has indices which are numerically the same as those of the plane. For example, in a cubic lattice the direction [1 1 0] is normal to the plane (1 1 0).

7.4. MICRO-METROLOGY OF CRYSTALS

7.4.1. General

Undoubtedly one of the most powerful tools for the quantitative study of materials' structure on the atomic or molecular scale is the x-ray spectrometer. Early workers in this particular field included von Laue, W. Friedrich and P. Knipping, but a tremendous advance was made in the understanding and development of the subject by Sir W. Bragg and his son W. L. Bragg (now, of course, Sir Lawrence).

7.4.2. X-ray Crystallography and Bragg's Law

The fundamental physical phenomenon brought into play in x-ray crystallography is that of wave interference. If two identical coincident wave systems (such as electromagnetic waves or pressure waves) arrive at a place absolutely in phase they will reinforce each other and produce a large intensity effect at that place. However, at a place where the two wave systems arrive separated by a phase angle $\Delta\theta = \pi$ (that is if they are half a wavelength apart in distance and half a period apart in time) they will cancel each other and there will be zero intensity (*see* Sub-section 1.3.1).

Thus if a series of identical wave trains of wavelength λ from a single source arrive at such positions that each of them has travelled further than the next by half a wavelength, or any odd number of half wavelengths $\left[\text{i.e. } \dfrac{1\lambda}{2}, \dfrac{3\lambda}{2}, \dfrac{5\lambda}{2}, \dfrac{7\lambda}{2}, \dfrac{9\lambda}{2} \cdots \dfrac{(2n+1)\lambda}{2} \right]$, they will result in zero intensity at those positions. On the other hand, at those positions where each of the wave trains has travelled further than the next by any whole number of wavelengths (i.e. 1λ, 2λ, 3λ, 4λ, ... $n\lambda$) the waves will reinforce each other and produce maxima of intensity—n is to be taken as an integer or 0 in the above.

The essence of Bragg's contribution is that the various planes of atoms in crystalline materials are considered to act as partially reflecting surfaces for x-radiation. Path differences therefore exist

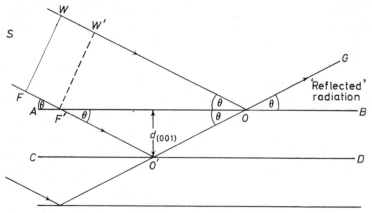

Figure 7.13. Diagram for derivation of Bragg's law

between rays reflected from one such plane and those reflected from adjacent parallel planes. Suppose, in *Figure 7.13* that the lines AB and CD represent adjacent (0 0 1) planes in a crystal having a primitive cubic lattice (it could be potassium chloride for example), and that these planes are separated by distance $d_{(0\,0\,1)}$.

Consider the rays WOG and $FO'G$ from S, a source of monochromatic x-radiation (*see* Sub-section 5.2.2), arriving at G, a radiation detector such as a Geiger-Müller tube (*see* Sub-section 4.2.3) or other suitable ionisation chamber. If WF represents an x-ray wave front then, by definition, it is perpendicular to both WO and FO': let $W'F'$ be parallel to WF.

It can be seen that the ray $FO'G$ has travelled further than the ray WOG by the amount $(F'O'O - W'O)$. Since, from the laws of optical reflection and from simple geometry, the angles θ are all equal, the geometry of *Figure 7.13* gives

$$F'O'O = 2F'O'$$
$$= 2d_{(0\,0\,1)} \frac{1}{\sin \theta}$$

and
$$W'O = F'O \cos \theta$$
$$= 2d_{(0\,0\,1)} \cot \theta \cos \theta$$
$$= 2d_{(0\,0\,1)} \frac{\cos^2 \theta}{\sin \theta}$$

The path difference between the two rays is thus given by

$$(F'O'O - W'O) = 2d_{(0\ 0\ 1)} \frac{1}{\sin \theta} (1 - \cos^2 \theta)$$

$$= 2d_{(0\ 0\ 1)} \frac{1}{\sin \theta} \sin^2 \theta$$

$$= 2d_{(0\ 0\ 1)} \sin \theta$$

If reflections from the planes *AB* and *CD* reinforce each other it follows that reflections from other (0 0 1) planes will also contribute to reinforcement, giving intense reflected radiation in the direction *OG*. Combining the above result with the previously stated requirement that for two wave trains to reinforce and produce maxima of intensity their path difference shall be an integral number of wavelengths, Bragg's law is obtained. This simply states that the condition for reflections of maximum intensity is

$$2d \sin \theta = n\lambda$$

where d is the interplanar separation, n an integer and λ the radiation wavelength.

7.4.3. X-ray Techniques

It should be apparent that if the (0 0 1) planes like *AB* and *CD*, in *Figure 7.14*, act as reflecting planes for x-rays, then so also will the (0 1 1) and the (0 1 2) planes represented by *KL*, *MN* and *PQ*, *RS* respectively, and all other such planes. These planes contain progressively fewer atoms per unit area and will consequently give rise to progressively weaker reflections. Also, their inter-planar

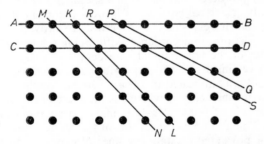

Figure 7.14. Various reflecting planes in x-ray crystallography

spacings $d_{(0\,1\,1)}$ and $d_{(0\,1\,2)}$ etc., become progressively less and must therefore be associated with progressively larger values of θ to meet the condition imposed by Bragg's law for a given x-ray wavelength.

The original Bragg method involved using a single crystal, and rotating both the crystal and the detector to obtain the various values of θ associated with the different inter-planar spacings $d_{(0\,0\,1)}$, $d_{(0\,1\,1)}$, $d_{(0\,1\,2)}$ etc. This is quite laborious, but on the other hand it has the merit that the results are straightforward to interpret. The method is illustrated schematically in *Figure 7.15*.

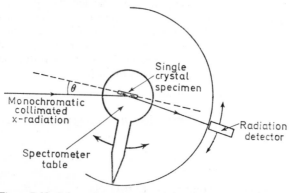

Figure 7.15. Schematic arrangement of Bragg x-ray spectrometer

A more widely used technique, which provides data for all the inter-planar spacings at one go, is the Debye-Scherrer powder method. A small quantity of the crystalline material to be observed is very finely powdered and placed in a specimen holder. Amongst the randomly orientated particles of powder there will always be a proportion (albeit very small) so placed as to provide maximum intensity reflections, in accordance with Bragg's law, for all the major planes of the lattice under consideration. These reflections will be in three dimensions and so provide conical surfaces of maximum intensity, as indicated in *Figure 7.16*, the semi-angle at the centre of each cone being 2θ where θ is the angle of the Bragg law.

The recording of the angles 4θ is made photographically on a film strip XY placed on a circular arc, at the centre of which is the specimen of powdered crystal. The lines of exposure on the film strip enable θ to be evaluated. For example, the distance S_{00}

between the lines marked (0 0 1) on the film strip in *Figure 7.16*, divided by the radius R of the arc of the film strip when in the spectrometer is four times the angle θ_{001} for the planes (0 0 1), and from this and Bragg's law $d_{(0\,0\,1)}$ is obtained.

It is almost entirely from the application of such techniques that the present knowledge of crystalline micro-structure stems.

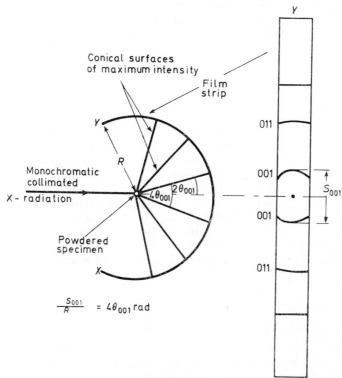

Figure 7.16. Schematic arrangement of Debye-Scherrer powder method x-ray spectrograph

7.5. ELECTRON DIFFRACTION

In the same way that x-radiations can be used to determine inter-planar spacings of crystals by use of Bragg's law, so also can electron beams. In Chapter 1, Sub-section 1.3.2 it was indicated that de Broglie's hypothesis of a wave property for a beam of electrons was valid (Davisson and Germer, 1927). A stream of electrons of velocity v has an effective wavelength λ given by

$$\lambda = \frac{h}{m_e v}$$

and such a stream of electrons considered in conjunction with Bragg's law enables inter-planar distances to be determined. Furthermore, much shorter wavelengths can readily be achieved with electron beams than with x-rays, and in this sense the electron diffraction technique is more powerful than the x-ray techniques.

7.6. CRYSTAL IMPERFECTIONS

7.6.1. General

It is very rare, except in the case of specially prepared specimens, to find any crystalline material exhibiting a completely perfect lattice structure over dimensions much in excess of a few hundreds of atomic spacings. Certain perfect crystals, for example of carbon, can be made in the form of long fibres and have recently led to the achievement of remarkably strong composite materials. The development potential in this field of Materials Science appears to be tremendous, but much work still remains to be done.

Various types of crystal imperfections exist and they are of great importance in relation to the physical and chemical behaviour of a material. These defects can be classified as follows.

7.6.2. Point Defects

These are associated with single lattice positions and are of several types.

(*a*) *Vacancy*—an unoccupied lattice site where an atom, ion or molecule would normally be expected.

(*b*) *Interstitial*—an additional atom situated between normal lattice planes.

(*c*) *Substitutional Impurity*—a lattice site occupied by an atom or ion of an element 'foreign' to the lattice.

(*d*) *Interstitial Impurity*—an atom or ion of a 'foreign' element situated between normal lattice planes.

The above are represented in *Figure 7.17*. In addition there are two other important defects, associated with ionic crystals, and these are

(*e*) *Schottky Defect*— a localised pair of vacancies, one at a cation site and the other at an anion site (*see* Sub-section 3.1.5).

(*f*) *Frenkel Defect*—a cation vacancy/interstitial pair, created by a cation moving from its normal lattice site to a nearby interstitial position.

The Schottky and Frenkel defects are illustrated in *Figure 7.18*.

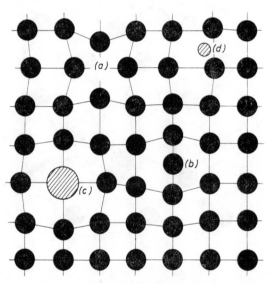

Figure 7.17. Simple point defects—(a) vacancy, (b) interstitial, (c) substitutional impurity, (d) interstitial impurity

7.6.3. Line Defects

Two parts of a crystal may move one with respect to the other and they are then said to have *slipped*. The interface of slip is called the *slip plane* and that which separates a slipped part of a lattice from the unslipped part is a *dislocation*. Dislocations are line defects and are of two kinds.

(*a*) *Edge dislocation*—this may be considered as the edge of an additional part-plane of atoms in the lattice and the dislocation

141

line vector lies along this edge. The presence of this dislocation, under shear stress, allows relatively easy movement (slip) to occur in a direction perpendicular to the dislocation line vector. The edge dislocation is represented by ⊥, and its relationship to slip is illustrated in *Figure 7.19*.

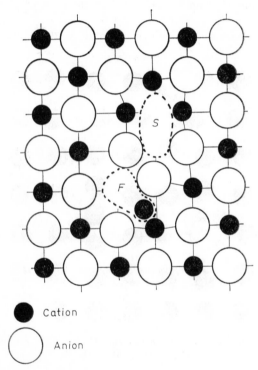

● Cation

○ Anion

Figure 7.18. Schottky (S) and Frenkel (F) point defects

(*b*) *Screw dislocation*—this is perhaps more difficult to visualise than the edge dislocation. Atoms can make small displacements, for example, from positions 1 to 2 and from 3 to 4 and the screw is represented by the path *A B C D E F G H*.... The dislocation *line vector* is along the 'core' of this screw and the element of slip is as indicated in *Figure 7.20*. This situation would be followed by the atom at *B* undertaking a 1 to 2 translation and that at *E* a 3 to 4 translation and so on until a complete unit of slip had occurred in the lattice. With screw dislocations, slip occurs in the direction of the dislocation line vector.

7.6.4. Surface Defects

These can be interfaces separating regions which, although having the same lattice orientation, are out of stacking sequence with one another; or they may be interfaces between regions in which the lattice orientations are different.

(a) *Stacking Fault*—this is the name given to the former of the above two cases and the dotted lines in *Figure 7.21* indicate such

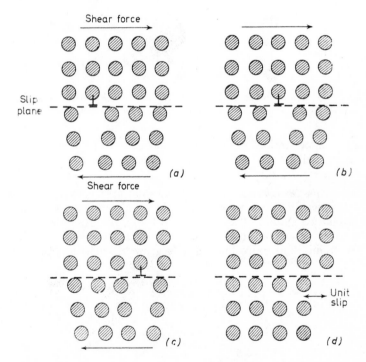

Figure 7.19. Edge dislocation, ⊥, and its progression under shear forces to produce unit slip

143

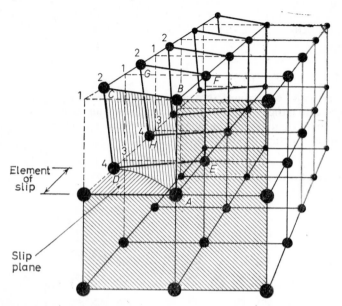

Figure 7.20. Screw dislocation ABCDEFGH.... showing element of slip and slip plane. Dislocation line vector is along 'core' of ABCDEFGH....

surface imperfections in two dimensions only. For the exact significance of the *A B C A B C A B....* and *A B A B A B A ....* stacking see Sub-section 6.3.3.

(*b*) *Grain Boundaries*—these are the interfaces between regions of a crystalline structure having different orientations of the same lattice pattern. They come into being, for example, during the solidification of a metal. Solidification starts at many 'nucleation' points and individual crystals grow at each of these, but

```
B B B B│C C C C│B B B B
A A A A│B B B B│A A A A
C C C C│A A A A│C C C C
B B B B│C C C C│B B B B
A A A A│B B B B│A A A A
C C C C│A A A A│C C C C
B B B B B̄ B̄ B̄ B̄ B B B B
A A A A A A A A A A A A
```

Figure 7.21. Stacking fault interfaces represented by broken lines

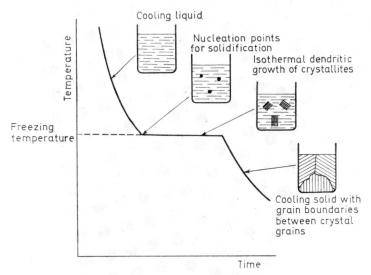

Figure 7.22. Idealised cooling curve illustrating development of grains and grain boundaries—for example, in a pure metal

each crystal will have a different orientation. At regions of contact between crystals, growth ceases, and when the entire quantity of material has solidified there will be as many grains as there were original nucleation points. This situation is illustrated schematically in *Figure 7.22*.

Grain boundaries are usually areas of gross mismatch between adjacent crystals, but a simple boundary in which the mismatch is not of a high order can take the form shown in *Figure 7.23*. This boundary is made up of a series of edge dislocations. It can be seen that a grain boundary offers a relatively large amount of space for the presence, or the movement, of impurity atoms. There is, in fact, a tendency for impurities to congregate at grain boundries and this can have a profound effect, especially upon the chemical properties of a material.

7.6.5. Gross Defects

Such large scale defects as complete voids, and large inclusions of foreign matter, can also occur in crystals.

145

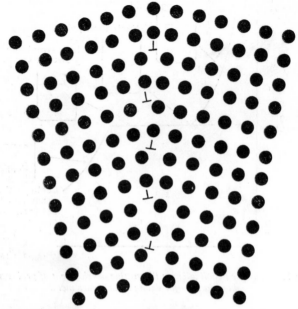

Figure 7.23. Simple grain boundary where the lattice mismatch is not of a high order

7.7. CRYSTAL IMPERFECTIONS RELATED TO MATERIAL BEHAVIOUR

The effects of the above types of defect on the behaviour—both physical and chemical—of materials is very great and can only be briefly indicated here. For example, the density of vacancies, whether simple or of the Schottky or Frenkel type, will greatly affect the rate of diffusion of a 'foreign' species through a lattice. This is because the process of diffusion is primarily a movement of the diffusing species from one otherwise vacant site to another.

The application of heat to a crystal not only causes an increase in vibrational energy of the lattice constituents, but also increases the number of vacancy defects and this results in an increase in the rate of diffusion for a given species. Interstitial diffusion can occur, but is limited to atoms which are very small compared with those of the main structure, and for such small atoms interstitial movement is comparatively easy anyway. It follows that lattice point

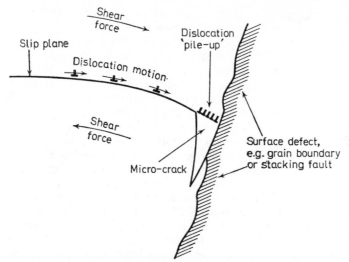

Figure 7.24. Crack initiation by dislocation 'pile-up' at surface defect

defects have little influence upon interstitial diffusion. Both types of diffusion, however, are greatly influenced by the number of line and more particularly, surface defects. For example, diffusion is much more rapid in a metal which has many small grains and therefore a greater area of surface defect than in one which has a few large grains.

The relatively small energy associated with the movement of edge and screw dislocations gives rise to easy slip between two sections of lattice, and largely accounts for the ductility ('workability') of certain materials, particularly metals. The production of unit slip by the motion of a single edge dislocation is illustrated in *Figure 7.19*.

Excessive stresses applied to a material may actually cause the generation of a series of dislocations whose movement can lead to a dislocation *pile-up* at any discontinuity in the lattice structure such as a stacking fault or a grain boundary. This pile-up, being unable to disperse because of the discontinuity, creates a micro-crack in the material, as shown *Figure 7.24*, and may well be the initiation of structural failure.

Grain boundaries and stacking faults both involve considerable lattice mismatch, and are regions to which impurities tend to migrate and then congregate. The concentration of impurity at grain boundaries often has a significant bearing upon the chemical behaviour of a material—for example it may cause a metal, or alloy, to become susceptible to intergranular corrosion, or, in the case of certain stainless steels, to 'weld decay'.

Substitutional impurities have an important significance in semi-conductor technology. For example, the deliberate addition, to covalent germanium, of a small proportion of substitutional atoms, whose size is nearly that of the germanium atoms, will not disrupt, but merely slightly distort, the covalent lattice pattern. Arsenic is one such impurity and it has five valence electrons. When the covalent bond is completed between each impurity arsenic atom and its four coordination germanium atoms (*see* Sub-section 6.3.2) one of the five arsenic valence electrons will be surplus, as shown in *Figure 7.25*. The energy required to remove this surplus electron from the proximity of its parent arsenic atom is quite small. Under the influence of a very small force it will move easily through the germanium lattice and thus contribute,

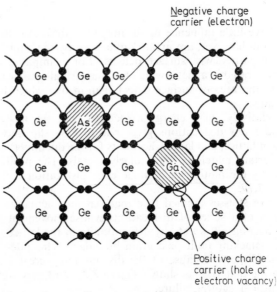

Negative charge carrier (electron)

Positive charge carrier (hole or electron vacancy)

Figure 7.25. Introduction of negative and positive charge carriers into covalent germanium by using substitutional impurity addition of Arsenic (As) and Gallium (Ga)

as a negative charge carrier, to electrical conduction. This induced or extrinsic, conductivity can be made to completely dominate the intrinsic semi-conduction of the material, in which case the impure germanium is referred to as n-type impurity semi-conductor.

Another element having appropriately sized atoms for addition to germanium is gallium. This has three valence electrons. Where a gallium atom is substitutionally situated in the germanium lattice the covalent bonding is short of one electron. The location for this missing electron is referred to as a *hole*, and this too is illustrated in *Figure 7.25*.

The amount of energy required for a neighbouring bond valence electron to transfer to, and thereby fill, a hole is not great, but in doing so, of course, it leaves a hole where it came from. This in turn can be filled by a further valence electron transfer and so on, providing a contribution to electrical conduction by this atom-to-atom 'switching' mechanism of valence electrons. From some points of view it is convenient to think of the movement of holes in one direction rather than the true switching movement of valence electrons in the opposite direction, and to treat the holes as positively charged entities which contribute in their own right to the conductivity of the germanium. This type of conduction is referred to as p-type impurity semi-conduction because the charge carriers, considered as holes, are positively charged.

Such n- and p-type impurity semi-conductors, and more particularly junctions between n- and p-regions in a crystal, are of great importance in modern electronics, such junctions being solid state *diodes*. A double junction in a crystal, comprising p-n-p- or n-p-n-regions, is a *transistor*, a physically small device which revolutionised the electronics industry, and which was developed from a theoretical knowledge of the structure of the materials involved.

7.8. QUESTIONS

1. Given that the density of pure crystalline NaCl is 2.165×10^3 kg m^{-3}, use the known value of Avogadro's number to determine the lattice constant [i.e. the spacing between cubic planes such as (0 0 1)] for this primitive cubic ionic crystal. Take the atomic weights for sodium and chlorine from *Table 3.1*.

2. Given that the (0 0 1) plane spacing in α-Fe is $2.866\,8 \times 10^{-10}$ m calculate a value for the (0 1 1) and for the (0 1 2) plane spacings. α-Fe has a body centred cubic structure.

3. The (0 0 1) planes of rock-salt (natural NaCl) are cleavage planes and are known to be separated by $2.819\,7 \times 10^{-10}$ m.

In a Bragg x-ray spectrometer, a beam of monochromatic x-radiation incident upon a single crystal of rock-salt gives successive maximum intensity reflections for glancing angles 5° 23′ and 10° 49′ with respect to a cleavage plane. Calculate a value for the radiation wavelength.

4. A fine, collimated, 5 keV electron beam is incident on a small sample of powdered graphite to produce electron diffraction in the manner of the Debye–Scherrer method. The interplanar spacing of the graphite structure is known to be 3.350×10^{-10} m. What is the apex angle of the cone representing the surface of first order Bragg reflection?

5. If x-rays were to be used in place of the electron beam of Question 4, what would be the energy of the x-radiation to provide the same angular result?

6. In a sample of n-type impurity germanium the number of substitutional arsenic impurity atoms in the germanium lattice is one in ten thousand. Calculate a value for the density of extrinsic charge carriers for this sample, taking the density of germanium as 5.46×10^3 kg m^{-3}. Use *Table 3.1* for atomic weights.

Chapter 8

Non-crystalline and Multiphase Solids

8.1. INTRODUCTION

The fundamental property which decides whether or not a material be classified as crystalline is long-range order in its micro-structure. Prior to the advent of x-ray methods in crystallography it was assumed that if a material did not exhibit such properties as a regular geometrical macro-structure, with specific angles between faces, and/or the ability to cleave along certain well defined planes, it was completely amorphous.

Based mainly upon the findings of x-ray and electron diffraction methods, short-range order, that is order over the range of a few atomic, or molecular, spacings, is now known to exist in most non-crystalline materials. Thus most so-called amorphous materials are not completely devoid of structural form, despite their name.

The difference between long-range and short-range order shows up particularly well in, for example, a comparison of Debye-Scherrer powder x-ray spectrographs. A crystalline material provides many sharply defined lines representing the high degree of long-range order possessed by such materials. On the other hand almost any non-crystalline material will produce one, or perhaps two, very diffuse zones of greater-than-average exposure of the recording film. Complete lack of order, that is true amorphism, should produce an absolutely smooth distribution of exposure with no preferred zones at all.

The short-range order which exists in non-crystalline materials can represent the structure of individual large molecules, or it may be associated with ordered groups of a few sub-units, such as ligancy polyhedra (*see* Sub-section 6.4.2).

8.2. NON-CRYSTALLINE ELEMENTS

Although it is possible to produce amorphous forms of a number of elements by invoking very large and rapid temperature changes, only the elements sulphur and selenium will produce non-crystalline structures when quenched from the molten state to normal ambient temperatures.

The bond in these elements is covalent and involves the formation of long-chain, or ring-type, molecules. These large molecules become very tangled in the disorder of the liquid (molten) state, so that very rapid cooling may allow insufficient time for disentanglement, resulting in a disordered solid. The sulphur and selenium type of bond is also mentioned in Sub-section 6.2.3, and is illustrated in *Figures 6.7* and *6.8*.

8.3. LONG-CHAIN MOLECULAR STRUCTURE

8.3.1. General

Mention has already been made of long-chain polymer molecules in Sub-section 6.4.3. These chains can be very long, comprising up to several million mers. They can also have substitutional side groups which may themselves be long chains, rings, or simply single atoms.

Both the great length of such chains and the presence of any side groups contribute to entanglement and to loose packing of the molecules which in turn promotes a more random solid state structure. It is energetically difficult for a random community of long, complicated molecules to reorientate into an ordered pattern. This despite the fact that, apart from the mechanical interlocking associated with their entanglement, the only forces binding most long chain molecules together are the comparatively weak van der Waals' forces.

Structurally, the simplest polymer molecule is the pure hydrocarbon polyethylene, C_nH_{2n+2}, where n may be of the order 10^4 to 10^6, and this molecule has already been mentioned in Sub-section 6.2.3. The polymer can exist in two possible forms: (*a*) linear, and (*b*) branching, shown in *Figure 8.1*.

In the solid state, though rarely completely crystalline, linear polyethylene usually exhibits a high degree of long-range order whilst such order is present to a far smaller degree in branching

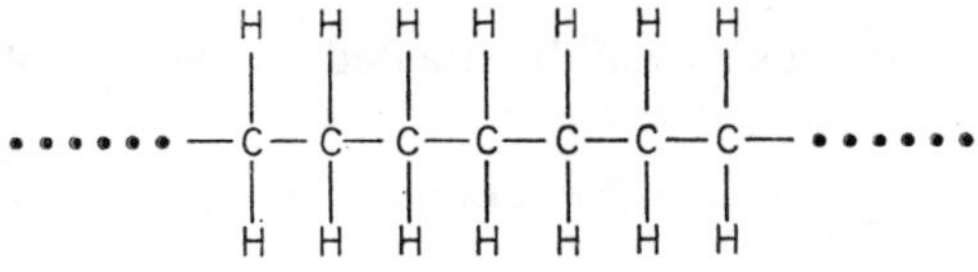

(a)

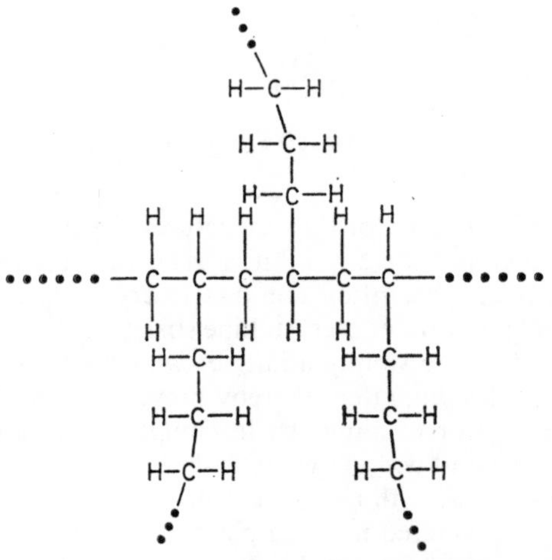

(b)

Figure 8.1. Polyethylene molecule: (a) linear, (b) branching

polyethylene because the very nature of its molecular structure encourages random packing. It is reasonable to assume that the disentanglement and ordering of branching structures will be energetically more difficult than for non-branching structures.

8.3.2. Vinyl Polymers

These comprise a repeating unit mer possessing a substitutional side group (SG) of appropriate valency to provide a covalent bond with a carbon atom of the main chain, as illustrated in *Figure 8.2.*

153

Such polymer structures may possess one of three possible arrangements:

(*i*) The side groups may regularly alternate from one side to the other of the molecular chain—termed *syndiotactic.*

(*ii*) The side groups may all be situated on one side of the molecular chain—termed *isotactic.*

(*iii*) The side groups may be randomly situated on the two sides of the chain—termed *atactic.*

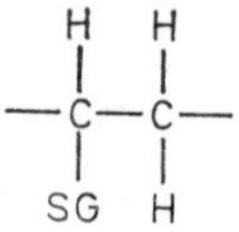

Figure 8.2. Substitutional side group (SG) in vinyl mer

Because of the regularity of their structures the syndiotactic and the isotactic vinyl polymers exhibit a relatively high degree of long-range order and are often completely crystalline. They can, however, be made to form non-crystalline structures by the addition of *plasticisers* which are simply additives which serve to keep the long chain molecules separated, thereby preventing crystallisation. Loss of plasticiser (for example, by leaching, or by vaporisation) can occur with age and will allow crystallisation to occur. This, of course, involves a loss, with age, of the property of plasticity, and this is a defect encountered in some plastic materials.

Atactic vinyl polymers, and especially those having large-size side groups, almost invariably produce non-crystalline structures. Examples are polyvinyl chloride (PVC) in which the side group is the single large atom, chlorine (large, that is, compared with the hydrogen and carbon atoms), and polystyrene, in which the side group is C_6H_5 (from the benzene ring structure C_6H_6—*see* Subsection 6.2.3). These two atactic vinyl polymers are represented in *Figure 8.3.*

8.3.3. Copolymers

Copolymers are made up of long molecular chains each consisting of two or more polymers. Such chains may be arranged with the individual mer units placed (*a*) randomly, (*b*) regularly alternating,

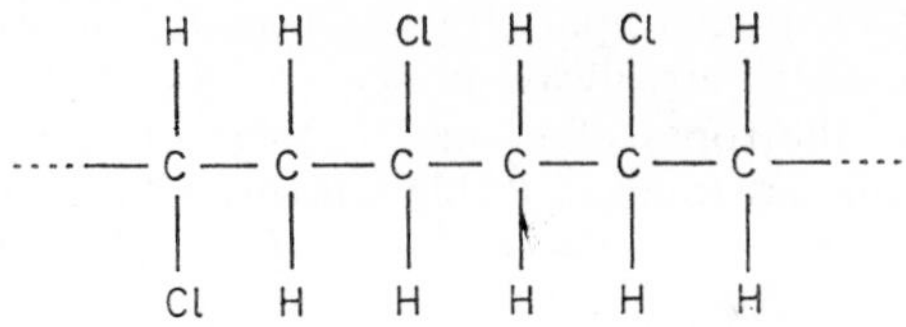

(a)

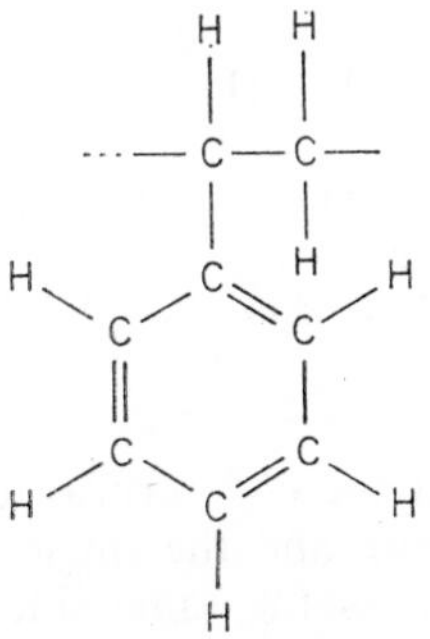

(b)

Figure 8.3(a) Random (atactic) distribution of Cl side group in polyvinyl chloride (PVC).

(b) The mer of the atactic polymer polystyrene

(c) in groups comprising alternating lengths of each polymer—a *block copolymer*, or (d) in groups such that a long chain of one polymer has side chains of the other—a *graft copolymer*.

8.3.4. Elastomers

Elastomers are polymers which possess a high degree of elasticity, that is, they obey Hooke's law for up to several hundred percent extension, (*see* Sub-section 1.1.6).

Elastomers are all non-crystalline at normal ambient temperatures. Their molecular chains are extremely long and very contorted, thus encouraging a high degree of entanglement and structural disorder. In addition, the molecular chains of elastomers are *cross linked*—that is, the chains are interconnected by strong bonds. This interconnection is provided either by single atoms, or by groups, forming primary bonds between the chains every few

155

hundred carbon atoms apart, in addition to the weak van der Waals' bonds which are always present.

One of the allotropes of natural rubber provides an excellent example of the elastomers. Rubber involves the methyl radical CH_3, having the structural formula

$$H-\overset{\displaystyle H}{\underset{\displaystyle H}{\overset{|}{\underset{|}{C}}}}-H,$$

as the side group.

The structure of the mer of the elastic allotrope of natural rubber, isoprene, is

$$-\overset{\displaystyle H}{\underset{\displaystyle H}{\overset{|}{\underset{|}{C}}}}\!\!-\!\!\overset{\displaystyle H-\overset{\displaystyle H}{\overset{|}{C}}-H}{\overset{|}{C}}\!\!=\!\!\overset{\displaystyle H}{\overset{|}{C}}\!\!-\!\!\overset{\displaystyle H}{\underset{\displaystyle H}{\overset{|}{\underset{|}{C}}}}\!\!-$$

This is referred to as the *cis** form of the polymer because the CH_3 substitutional group and the single H on either side of the double bond are both on the same side of the chain. Such *cis* formation is primarily responsible for the contorted form of elastomer molecules which is a major factor in providing the property of elasticity.

Another allotropic form of natural rubber, gutta-percha, is constitutionally the same as the elastomeric allotrope but is structurally different in the following particular. The mer in this case has the CH_3 side group and the single H on either side of the double bond on opposite sides of the chain. This arrangement is referred to as the *trans* form of the polymer, and the structure of the isoprene *trans*-mer is

$$-\overset{\displaystyle H}{\underset{\displaystyle H}{\overset{|}{\underset{|}{C}}}}\!\!-\!\!\overset{\displaystyle H-\overset{\displaystyle H}{\overset{|}{C}}-H}{\overset{|}{C}}\!\!=\!\!\overset{\displaystyle H}{\overset{|}{C}}\!\!-\!\!\overset{\displaystyle H}{\underset{\displaystyle H}{\overset{|}{\underset{|}{C}}}}\!\!-$$

The *trans*-polyisoprene molecule has a more balanced structure and is therefore less contorted than the *cis* molecule. The overall solid state structure of the *trans* polymer is therefore less disordered,

---

* cis—Latin 'on this side'; trans—Latin 'across'.

156

and the material can rightly be expected to tend to crystallisation and therefore to be more rigid than the *cis* form.

cis-Polyisoprene is simply a viscous liquid if it is not cross-linked and will, in fact, crystallise at around 273 K (0° C); *trans*-polyisoprene, on the other hand, is crystalline below about 333 K (60 °C). The most widely used cross-linkage is provided, quite easily, by the addition of a small percentage of sulphur (S) to the rubber. The cross-link is achieved by the breaking of some of the

double bonds and the creation of a —C—S—C— link, as illustrated

in *Figure 8.4.*

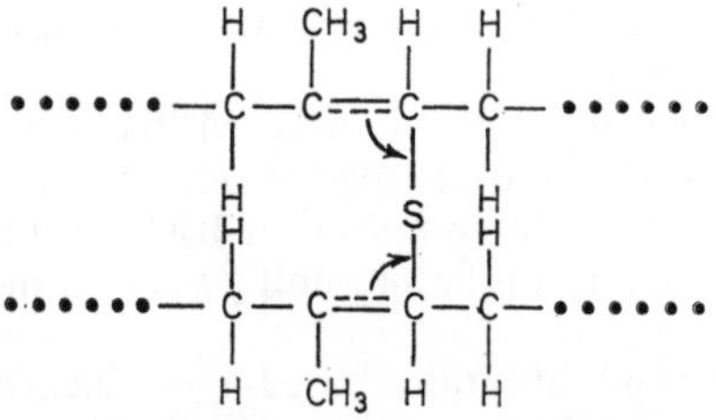

Figure 8.4. The sulphur cross-link in cis-polyisoprene

As the degree of cross-linkage increases in such a material, its structure will become more and more rigid. Thus it is possible, from natural rubber, to provide materials with characteristics ranging from those of the highly elastic 'india-rubber' to those of the virtually rigid 'ebonite'.

The process in which sulphur is made to form cross-links is called *vulcanisation* and involves heating the rubber with sulphur. The heat energy achieves the breakage of a proportion of the double bonds, then, on cooling, cross-linkage occurs since the sulphur cross-link bond is energetically preferred to the re-establishment of a double bond.

8.4. GLASSY, OR VITREOUS STRUCTURES

Glasses are produced by solidifying a melt of suitable constituents so rapidly that there is insufficient time for significant atomic or molecular diffusion to occur during the transition. The structural disorder associated with the liquid state is thus 'locked into' the solid state. The term *glass* refers, not so much to the components of the material, but rather to its structural form.

Glasses comprise three-dimensional, disordered, open (i.e. not close-packed) networks of various sub-units. In commercial glasses these sub-units are mainly oxide ligancy polyhedra of silicon and/or boron. The former are $(SiO_4)^{4-}$ tetrahedra (*see* Sub-section 6.4.5) which join corner-to-corner to provide the compound silica (SiO_2), whilst boron oxide, which has a cation to anion radius ratio of $\frac{0.23}{1.32} = 0.175$, is expected to form a triangular ligancy polyhedron $(BO_3)^{3-}$ (*see* Sub-section 6.4.2). This trigonal borate sub-unit is not, in fact, completely planar, the boron ion being slightly out of the plane of the three oxygen ions. In a similar manner to that of the silicon oxide tetrahedra these trigonal borate sub-units join corner-to-corner and again provide an open and structurally disordered network.

Since the corner-to-corner joining of the $(BO_3)^{3-}$ sub-units involves the sharing of each oxygen ion between two trigonal sub-units, the true borate composition B_2O_3 results because each of the three oxygen ions of the sub-unit gives only half its allegiance to one boron ion.

Most glasses are modifications based upon silica and, in commercial production, additives such as soda (Na_2O) and lime (CaO) may be used to break up the pure silica networks, thereby promoting closer packing of the structure and encouraging some crystallisation. These additives also lower the temperature at which the glass can be formed and are termed *modifiers*. Other additives have the reverse effect and actively encourage structural disorder, these are termed *glass formers*, examples are Be_2O_3 and GeO_2.

Structurally, glasses are often described as super-cooled liquids because the structural disorder associated with the liquid state remains throughout the transition to the solid state due to very rapid cooling. Because the bonds of glass structure have varying energies they do not all disrupt at a particular temperature when solid glass is heated, but rather they progressively break down as the temperature rises. Thus the material slowly becomes less rigid, then increasingly plastic until it becomes completely viscous. The *glass transition temperature*, T_g, is defined as the mid-temperature between the glass being brittle and being viscous—a rather vague and sometimes difficult quantity to evaluate.

Many special purpose glasses are manufactured, and involve various metals substitutionally replacing silicon in the vitreous structure. Typical metals for such purposes are lead, barium and calcium.

8.5. MULTI-PHASE SOLIDS

8.5.1. General

Different components in a material are often referred to as different *phases*, and the significance of this term will be elaborated in Chapter 9. This section deals with materials which comprise aggregates of several components, or phases, in the solid state.

Probably the most commonplace of the multi-phase solids are the natural rocks. Evidence of the importance of these materials is the extent to which they are used as building materials, and their obvious properties of load-bearing and resistance to decay, quite apart from their attractive appearance.

An example of a magnificent building rock is granite. This is formed by the cooling of *magma*, a natural molten silicon and oxygen-rich mixture of many elements, which exists deep in the earth's interior. Granite comprises several crystalline minerals (phases), but principally mica and quartz (*see* Sub-section 6.4.5) in a matrix of the major component felspar (another complex silicate).

If, during the original formation of granite, the magma cooled very slowly there would be ample time for atomic, molecular and particle diffusion to occur and hence for small units (crystallites) of individual minerals to form and then to coalesce into large domains. This would give rise to a coarse structured rock, possibly containing large deposits of individual minerals. On the other hand more rapid cooling would produce a finer-structured formation (e.g. rhyolite). Extremely rapid cooling of the magma would prevent much atomic or molecular diffusion and so there could be little or no formation of individual crystals; the material resulting would have little or no long-range order and would therefore be glassy. Obsidian and opal are two such natural glasses.

8.5.2. Ceramics

Strictly speaking, the term 'ceramics' refers to those materials produced by the pottery industry. They include industrial, as well as domestic, earthenwares, chinas, tilings and bricks. These materials have certain properties not unlike those of some natural rocks. They are manufactured from the natural complex aluminium silicates known as *clays*. The major component comprises a glassy matrix formed during the firing stage of manufacture. This glassy phase bonds together grains of sand and particles of other siliceous constituents of the original clay which have not reacted to produce the glassy phase.

Ceramics have good load bearing properties (good compressional strength) but they are weak in tension and, being largely glassy, are very brittle. One of their most useful characteristics is the ability to withstand very high temperatures.

8.5.3. Concrete

This multi-phase solid comprises the many phases represented by selected and graded stones, gravels and sands which are collectively known as *ballast*. The bonding matrix is of *Portland cement* which is manufactured from natural limestones and shales (the latter being clays which have compacted under pressure to form rocks). In manufacture these are crushed, dried, ground small, subjected to very high temperature, then cooled to form a glassy clinker. This in turn is crushed, has some gypsum added, and is finally very finely ground to form the familiar grey powder called cement.

When cement is mixed with water the structure is *colloidal* (the solid particles of cement are so tiny that they remain in suspension in the liquid with no tendency to settle). The setting process involves the formation of a three dimensional, but disordered, network of bonds involving water molecules and the various silicates of the cement. Water which is surplus to bonding is driven off by evaporation. The final structure is very rigid and behaves as a single, non-crystalline (amorphous) phase.

The water involved in the setting of the cement should not be thought of in terms of its 'wetness' or even as bulk liquid but rather as individual hydrogen-oxide molecules which are essential links in the solid state bonding between the various silicates.

8.5.4. Cermets

Cermets, as the name suggests, are mixtures of ceramic and metallic materials. They can range from particles of ceramic material embedded in a matrix of metal, to the opposite end of the scale involving metallic filaments, platelets or other sub-units embedded in a matrix of ceramic. The potential of these materials lies in possibilities of combining the desirable properties of high melting point and great hardness of the ceramics with such metallic properties as good thermal and electrical conduction. Much research and development work is in progress in this field.

8.5.5. Timber

This multi-phase material, of course, originates from living matter and is therefore, by definition, *organic*. Timber consists of a tough fibrous cellular structure comprising bundles of long-chain molecules of *cellulose*. These molecules are made up of the elements hydrogen, carbon and oxygen having the constitutional formula $(C_6H_{10}O_5)_n$ where n is a large integer. The molecular bundles lie in spirals and are bound together to form the cell walls by a non-crystalline phase. The material of this phase is known as *lignin*, which also serves to bond the cells together. The spiral formation of the molecular bundles accounts, to a large extent, for the toughness of timber and thus for its great usefulness as a constructional material.

Chapter 9

Phases, Equilibrium Diagrams and Alloys

9.1. PHASES

Differences in constitution or in the structural arrangement of the constituent atoms, ions or molecules of materials give rise to different phases. The simplest concept of phase is probably represented by the three major phases, or states, of matter, *solid, liquid and vapour*, as instanced by *ice, water* and *steam* in the case of the hydrogen-oxygen compound water. These three phases comprise different structural arrangements of the constituents.

All elements, and most inorganic compound materials, are believed capable of assuming any of the above three major phases under appropriate conditions of temperature and pressure. For example, even helium (He), which is usually associated with the vapour phase, and occasionally with the liquid phase, can be made to assume the solid state, but only with considerable difficulty. Because of its extra-nuclear electronic structure (*see* Section 3.4 and *Table 3.1*) it would not be expected to achieve any of the bonds normally associated with the solid state (ionic, covalent or metallic). Helium liquefies at a temperature of about 4 K (in other words about 4° C above the absolute zero of temperature) and does not assume the solid state until the temperature is reduced to less than 1 K (1° C above absolute zero). At this very low temperature the constituent atoms have so little vibrational energy that they become sufficiently closely packed (*see* Sub-section 6.1.1) for the weak van der Waals' forces to induce the solid state.

The variables involved in producing different phases in materials are as follows:

(*a*) Composition of the material—this refers both to the number of components (which may be elements or chemical compounds, but not mixtures), and to their relative abundance.

(*b*) Temperature.

(*c*) Pressure.

9.2. THE GIBBS PHASE RULE

For a given system the thermodynamically derived Gibbs phase rule gives the relationship between the number of phases which can co-exist in equilibrium (*P*), the number of components that the system comprises (*C*) and the number of the above three variables which can change independently without changing *P*. This last is referred to as the number of degrees of freedom (*F*). The Gibbs phase rule is

$$P + F = C + 2$$

9.3. EQUILIBRIUM DIAGRAMS

Unary and binary equilibrium diagrams comprise a two-dimensional field on which any point defines each of two variables, the third variable being constant in value for a particular diagram. They are constructed to represent conditions under which the variables change so slowly that the system is always in equilibrium (i.e. completely stable). They take no account of such rapid changes as are produced, for example, by quenching.

Unary diagrams deal with a single component (*C* = 1 for the Gibbs phase rule), so the two variables plotted are temperature and pressure (the third variable, composition is constant by the very definition of a unary system).

Figure 9.1 is a schematic presentation of a unary equilibrium diagram on which are shown the three phases *solid, liquid* and *vapour*. These phases are separated by lines which are the loci of points of equilibrium between the phases. Also indicated, by arrows, are several possible phase transitions. Useful terms describing these phase transitions are contained in the key.

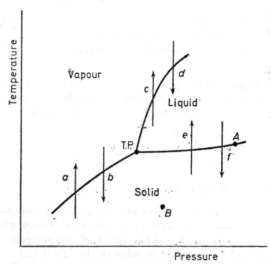

Figure 9.1. Schematic unary phase diagram

a *Sublimation*	d *Condensation*
b *Vapour deposition*	e *Fusion or Melting*
c *Vaporisation*	f *Solidification or freezing*

T. P. The triple point at
which all three phases
are in equilibrium

The phase changes shown are *isobaric*, that is they each occur without any change of pressure. It should be remembered that *isothermal* (constant temperature) changes can also occur and so can changes which are neither isobaric nor isothermal, and these last are the most common in practice.

Consider the phase rule applied to the system represented by *Figure 9.1.*

(1) At T.P.

$$P+F = C+2$$
$$P = 3, \text{ and } C = 1$$

so
$$3+F = 1+2$$

and
$$F = 0$$

In other words none of the variables can change without altering the number of phases in equilibrium.

164

(2) At point A

$$P + F = C + 2$$
$$P = 2, \text{ and } C = 1$$

so

$$2 + F = 1 + 2$$

and

$$F = 1$$

In this case one variable can change independently (necessitating a dependent change in the other) without altering the number of phases in equilibrium.

(3) At point B

$$P + F = C + 2$$
$$P = 1, \text{ and } C = 1$$

so

$$1 + F = 1 + 2$$

and

$$F = 2$$

and the two variables can change independently without altering the number of phases in equilibrium.

Figure 9.2 is an approximate unary diagram for pure iron (Fe) in which it is seen that there are three separate phases in the solid state. These are classified alpha (α), gamma (γ) and delta (δ) phases, and represent different structural arrangements of the atoms (allotropes). The α and δ-phases are BCC and the γ-phase FCC, (*see* Section 9.1). The α to γ transition temperature is also known as the *Curie point* and is of interest because apart from the structural change which involves a change of volume, the material changes its magnetic properties (from the everyday point of view it changes from being magnetic to being non-magnetic*).

Binary diagrams deal with two components so that $C = 2$ in applying the Gibbs phase rule. In such diagrams the variables to be plotted may be chosen from the following pairs.

(1) (*a*) Relative abundance of the two components
(*b*) Pressure
} at fixed temperature.

(2) (*a*) Relative abundance of the two components
(*b*) Temperature
} at fixed pressure.

(3) (*a*) Pressure
(*b*) Temperature
} at fixed proportions of the two components.

* Strictly it changes from being *ferromagnetic* to *paramagnetic*.

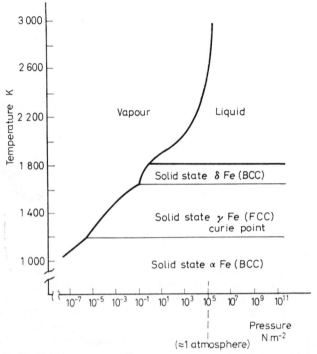

Figure 9.2. Approximate unary diagram for pure iron

Now although all the above are equally important from many points of view, in practical engineering the relationship (2) is usually of greatest general interest. Therefore most binary diagrams used in Materials Science comprise Temperature as ordinate versus Relative Abundance of Components as abscissa, plotted for a fixed pressure; but the other types of binary diagram should not be completely overlooked.

In relationship (2), since one degree of freedom has been eliminated by specifying a fixed pressure, the Gibbs phase rule has to be modified to the form

$$P+F = C+1$$

when using such binary diagrams.

The binary diagrams to be dealt with in this book are all as relationship (2) above and are related entirely to metal systems. They are discussed further in Section 9.5.

166

Ternary and *quaternary* equilibrium diagrams are concerned with three and four-component systems respectively and are obviously very important in modern Materials Science—particularly metallurgy—but they are beyond the scope of this book.

9.4. ALLOYING

9.4.1. General

Alloys comprise a metallic principal component with added alloying materials which are usually metals, but may be non-metals (for example carbon), or metallic compounds. This topic is dealt with here exclusively from the metallurgical standpoint and is restricted to binary systems.

In the liquid phase, practically all metals are capable of atomic mixing to the extent of producing a completely homogeneous material, in other words they are *miscible*. But in the solid state this is certainly not always so and the components may (*a*) be completely miscible, (*b*) be completely immiscible, (*c*) be partially miscible, or (*d*) form chemical compounds of the components.

As an example of (*d*), the intermetallic chemical compound Mg_2Pb occurs in the magnesium-lead binary system at about 19% magnesium and 81% lead by weight. Mg_2Pb is not an alloy but is simply another component which forms part of the system.

In the solid state, miscibility implies one component being completely dissolved in the other (not chemically combined with it) forming a *solid solution* which is a single phase. Complete immiscibility means that the two components simply solidify as themselves, producing, for example, mixed crystallites or even grains of the two distinct components or phases.

In the case of two partially miscible components A and B, the initial solid solution of B in A is usually referred to as phase α, and the initial solid solution of A in B as phase β. Solid solutions of different structure from α and β, or separated from phases α and β by regions of immiscibility, or by intermetallic chemical compounds, are termed intermediate phases and may be designated γ, δ, ε-phases and so on.

The metallic bond is such that it allows alloying constituents to enter a lattice in one of two ways, either substitutionally or interstitially (*see also* Sub-section 7.5.2). The latter is almost entirely restricted to very small size atoms such as those of hydrogen, boron, carbon and nitrogen.

167

Alloying additions are made with the definite objective of induc-ing, or improving, specific physical or chemical properties, but rarely are these properties gained without some sacrifice. For example, copper added to aluminium very greatly increases its strength to weight ratio (Al with about 4% Cu by weight is 'Dur-alumin'). But whereas pure aluminium has a great resistance to many forms of chemical attack because of the protective properties of its natural oxide Al_2O_3, the presence of copper considerably reduces the protection afforded by this oxide (probably by render-ing it porous). Strength is gained here at the expense of chemical resistance. As another example, chromium is added to irons and steels specifically to induce corrosion resistance (anything above about 15% by weight of chromium added to irons or steels provides what are called 'stainless' alloys). But the financial cost of such alloys is so great compared with ordinary steels that their use is severely restricted.

9.4.2. Guiding Rules for Alloying

It is found by experience that if an alloying additive is chosen simply at random it is more likely to produce undesirable forms of micro-structure than a useful homogeneous solid solution. The late Professor William Hume-Rothery developed a set of general rules applicable to the formation of substitutional solid solutions which are of considerable practical help in the formulation of binary alloys. These rules are essentially as follows:

(1) *Relative size rule*—If the atomic sizes of the two components differ by less than 15% then the size factor is favourable, and at least 10% solubility of one component in the other can be expected if other factors are favourable.

(2) *Chemical affinity rule*—The greater the chemical affinity* between the two components the more restricted will be the degree of solid solubility.

(3) *Relative valency rule*—If the number of valence electrons of the two components differs, alloying which tends to increase the average number of valence electrons per atom will be favoured more than alloying which tends to decrease the average number of valence electrons per atom. In other words the component with the smaller number of valence electrons will dissolve the component

* Chemical Affinity—the tendency for chemical combination to occur and produce a compound.

168

with the greater number better than the other way round. For example copper (normal* valency 1) dissolves zinc (valency 2) better than vice versa and nickel (normal valency 2) dissolves chromium (normal valency 3) better than vice versa.

(4) *Lattice type rule*—Solid solubility over large ranges of relative abundance of the two components is only possible for components having the same lattice structure. For example copper and nickel are both structurally FCC, and they exhibit solid solubility over all values of relative abundance.

The above rules, which are to some extent common sense, enable a qualitative idea to be gained of whether a possible alloy will involve complete solid miscibility, complete solid immiscibility or partial solid miscibility.

For interstitial solid solutions, the atoms of the added component must be of small diameter if they are to occupy interstitial positions without completely altering the structure of the principal component and producing another phase. In metallurgy, interstitial solid solutions are usually confined to additions of the elements hydrogen, boron, carbon and nitrogen, but even such small atoms placed interstitially must cause some distortion of the metal lattice. Since, for non-plastic materials, resistance to distortion is proportional to the degree of distortion (within the elastic limit, of course), it is not unreasonable to deduce that such interstitial components will cause hardening (hardness being usually defined in terms of resistance of the material to distortion, as measured, for example, by the indentation produced by a diamond point under specified conditions of loading).

9.5. BINARY DIAGRAMS

9.5.1. *General*

In *Figure 9.2.* it is seen that the phase transition from liquid to solid iron at around one atmosphere pressure occurs at a single temperature. A temperature/time cooling curve would be as *Figure 7.22*, which also indicates schematically some features of the macro-structure of the material at various temperatures.

For binary systems an isothermal liquid-to-solid phase transition

---

* The 'normal' valency is that associated with the lowest energy state of the atom.

is very rarely the case. Not unexpectedly, solidification proceeds over a range of temperature—the *freezing range*—which lies somewhere between the freezing temperatures of the two components.

9.5.2. Binary System of Complete Solid Miscibility

In this case the equilibrium diagram has a pair of lines, the *liquidus* and the *solidus*, whose vertical separation represents the freezing range for various proportions of the two components, as in *Figure 9.3*.

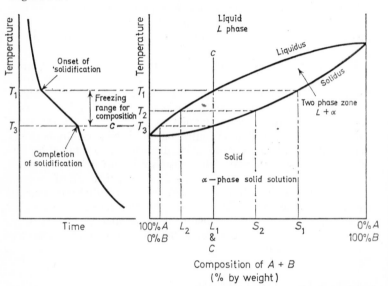

Figure 9.3. *Binary equilibrium diagram for components A and B exhibiting complete solid state miscibility. Also, cooling curve for composition C (about 70%A+30%B)*

Using *Figure 9.3* and considering a material having composition C (about 70% A and 30% B), the first solid specks appear at temperature T_1, are of composition S_1, and are in equilibrium with liquid of composition L_1. These specks are of higher B content than that represented by C, with the result that the remaining liquid is slightly depleted of B. As cooling, and solidification, proceeds (very slowly in order to maintain equilibrium conditions), at some intermediate temperature T_2 the solid (now in the form of

170

dendrites*) will have average composition S_2 and will be in equilibrium with liquid of composition L_2. The process of diffusion, which is comparatively easy at high temperatures, allows the necessary continuous change in the overall composition of the solid dendrites.

With further cooling and solidification the composition of the solid continues to follow the solidus and the corresponding liquid composition is always represented by the position on the liquidus horizontally opposite the point on the solidus. Final solidification occurs at the temperature T_3 at which the vertical representing the overall composition of the material C cuts the solidus. The final structure of the solid is granular, each of the original nucleation points for solidification developing into one grain. The interfaces between grains are, of course, grain boundaries which have already been discussed in Sub-section 7.6.4.

Horizontal lines joining the liquidus and solidus are called *tie-lines* and provide information not only on the composition of the solid and liquid in equilibrium (as S_2 and L_2 at temperature T_2 in *Figure 9.3*) but also on the relative quantities of solid and liquid that are in equilibrium at a particular temperature. This latter information is obtained by means of the *lever principle* which treats the tie-line as a simple lever having a fulcrum at the position representing the average composition C of the material.

If, at temperature T, W_L is the percentage weight of the material in the liquid phase and W_S is the percentage by weight in the solid phase, then

$$W_L \times l = W_S \times s \quad \text{[by the principle of the lever]}$$

and

$$W_L + W_S = 100\%$$

two simultaneous equations which allow the solution of two unknowns. This should be made clear by reference to *Figure 9.4*.

9.5.3. Binary System of Complete Solid Immiscibility

The equilibrium diagram for this system—the *simple eutectic* system—is shown in *Figure 9.5*. It represents two components A and B which are completely immiscible in the solid state. Taking the composition marked C, the liquid cools until the liquidus is

---

* *dendrites*—skeleton crystals of 'tree-like' (i.e. branching) form.

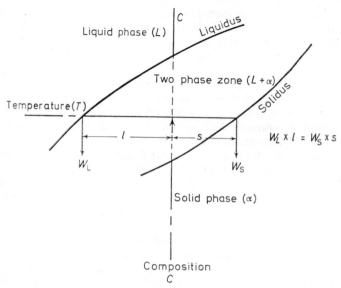

Figure 9.4. The lever principle for weights (strictly masses) of phases in equilibrium

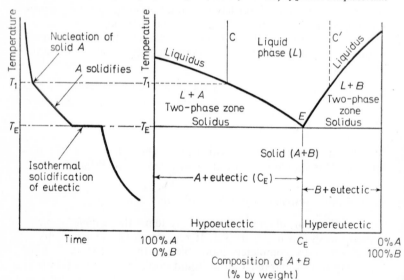

Figure 9.5. Simple eutectic equilibrium diagram for completely immiscible compo-nents A and B. Also, cooling curve for hypoeutectic composition C

172

reached. At this temperature, T_1, specks of the solid A begin to form. This leaves the liquid depleted of A. On further cooling, the remaining liquid composition follows the liquidus as more and more of A appears as solid, until the point E is reached. E is the *eutectic point*, T_E the *eutectic temperature* and C_E the *eutectic composition*. At E the temperature remains constant until all the remaining liquid has transformed to the solid state, which has the eutectic composition. The solid eutectic is an *intimate mixture* of A and B, each in the form of particles, platelets or rods, but is *not a solid solution*.

The final solid of composition C therefore comprises primary solid A (often single grains) embedded in a solid matrix of eutectic. In a similar fashion the final solid composition of C' would be primary particles of B in a solid matrix of eutectic. C is said to have a *hypoeutectic* composition whereas C' has a *hypereutectic* composition.

9.5.4. Binary System of Partial Solid Miscibility

There are two systems involving partial solid miscibility and these are the *eutectic* and *peritectic* systems.

The partial miscibility eutectic system is represented in *Figure 9.6*, in which there are seen to be three two-phase regions, $L+\alpha$, $L+\beta$, and $\alpha+\beta$.

Consider the hypereutectic composition C in *Figure 9.6*. At T_1 the first solid to appear comprises solid specks of β-phase (solid solution) having composition $S_{\beta I}$. Then, as cooling proceeds, the β specks grow and become, by diffusion, progressively richer in A according to the solidus, until the eutectic temperature T_E is reached, at which the β-phase has composition $S_{\beta E}$. At this temperature the remaining liquid, which is of eutectic composition C_E, is transformed isothermally to the two-phase solid eutectic, $(\alpha+\beta)_E$, and the proportions of α and β in the eutectic can be obtained from the lever principle, the fulcrum being at C_E.

When isothermal solidification has completed at T_E, further cooling necessitates a modification of the β-phase composition, by diffusion, to $S_{\beta F}$ at the final temperature T_F. At this temperature the eutectic comprises an intimate mixture of α phase of composition $S_{\alpha F}$ and β-phase of composition $S_{\beta F}$. The α and β constituents of the eutectic have changed composition slightly, following the *solvus** lines, but the overall composition in terms of the relative

---

* Solvus lines separate solid state phases

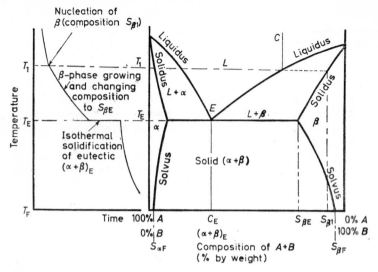

Figure 9.6. Partial miscibility eutectic system for components A and B. Also, cooling curve for hypereutectic composition C

abundance of the components A and B remains constant as represented by C_E.

The partial miscibility peritectic system is illustrated in *Figure 9.7*, and it is seen that, as in the eutectic system, there are three two-phase regions, $L+\alpha$, $L+\beta$ and $\alpha+\beta$. Consider an original melt of composition C which is allowed to cool slowly to maintain equilibrium conditions. The first solid specks appear at T_1 and are of β-phase having composition $S_{\beta I}$, then as cooling proceeds the solid β specks grow into crystals which become, by diffusion, progressively richer in A, whilst the remaining liquid becomes depleted of B. At the *peritectic temperature*, T_P, further heat loss produces no temperature change until solidification is complete. The remaining liquid of composition L_P and solid β of composition $S_{\beta P}$ react together to produce solid α of the *peritectic composition* C_P.

When this isothermal peritectic reaction at T_P has reached completion, further reduction of temperature involves solid state reactions in which the α composition follows the solvus line, which is only possible by the precipitation of solid β. The relative abundance and the composition of the α and β-phases in the $\alpha+\beta$ two-phase region can be obtained using the lever principle. The final

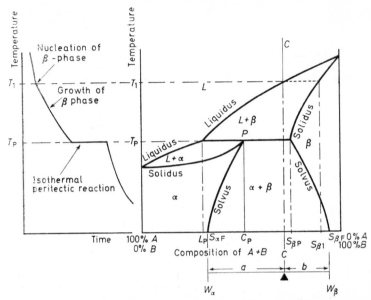

Figure 9.7. Partial miscibility peritectic system. Also, cooling curve for composition C and lever diagram for final proportions of α and β-phases

overall composition in terms of A and B is, of course, C, but the α and β proportions, by weight are given by

$$W_\alpha \times a = W_\beta \times b$$

and

$$W_\alpha + W_\beta = 100\%$$

the compositions of the two phases being $S_{\beta F}$ and $S_{\alpha F}$.

9.6. SOLID STATE TRANSFORMATION REACTIONS

9.6.1. General

Reactions similar to those just discussed can, in fact, occur in the solid state, involving changes from one solid state phase to another via a solid state two-phase region. Solid state reactions of the eutectic type are termed *eutectoid* and those of the peritectic type are termed *peritectoid*.

The iron–carbon system involves various types of solid state reaction including one eutectoid, and is shown in *Figure 9.8*.

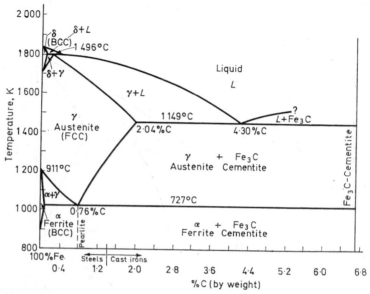

Figure 9.8. Equilibrium diagram of Fe-Fe₃C system (The values quoted are from J. S. Kirkaldy, Aspects of Modern Ferrous Metallurgy, Blackie, London, 1964)

This particular system is really only of interest between Fe 100% and Fe 93.3%, C 6.7%, this last composition representing the chemical compound iron carbide Fe_3C (the atomic weight of Fe is 55.85 and that of C is 12.01, therefore the total weight of the Fe_3C molecule is

$$3 \times 55.85 + 12.01 = 179.56$$

and of this the percentage weight of the carbon in the molecule is

$$\frac{12.01}{179.56} \times 100\% = 6.69\%)$$

The compound Fe_3C has the name *cementite* and like many other metallic carbides it is extremely hard. The two components of the most interesting part of the iron–carbon phase equilibrium diagram are thus elemental iron (Fe), and the compound cementite (Fe_3C).

Because of the great practical importance of the binary iron carbon alloys and the depth to which they have been studied—they are the basis of all steels and cast irons—the various phases of the system have come to have particular names, and these are also shown in *Figure 9.8*.

9.6.2. Non-equilibrium Cooling

Although the diagrams discussed in this chapter relate specifically to equilibrium conditions, their application is by no means restricted to such conditions. For example, suppose a billet of plain carbon steel of 0.4% carbon is raised to, and held at, a temperature within the γ-phase (say about 1 300 K or roughly 1 000° C) for a sufficient time to allow it all to attain this particular phase structure (FCC solid solution called *austenite*). If it is then *very suddenly* cooled (quenched)—down to about 320 K, i.e. roughly 50° C—there will be no time for any diffusion to occur. The resulting solid will be a supersaturated solution of C in Fe because at around 50° C it can be seen that the maximum solubility of C in Fe, represented by the α-phase solvus line of the equilibrium diagram, is very slight. The crystal lattice will therefore be extremely distorted. The amount of lattice distortion will naturally depend upon the carbon content, but the supersaturated solid solution must be very hard and brittle (*see* Sub-section 9.4.2, last paragraph). This very hard supersaturated solid solution of carbon in iron is called *martensite*.

The hardening of steel, by quenching to produce martensite, is a preliminary to the process of *tempering*, which involves heating the martensitic steel to an appropriate temperature thus allowing a controlled amount of diffusion. The consequence of such diffusion is the precipitation of some of the carbon, which is in excess of solid solubility, in the form of the compound cementite, producing an appropriate amount of softening and toughening in the steel.

The understanding of such processes is made easier by reference to, and appreciation of, the equilibrium diagrams. Further discussion of this subject rightly belongs in a treatise on Metallurgy, being beyond the scope of this introductory monograph on physical principles of Materials Science.

9.7. QUESTION

1. Alpha-phase Fe (BCC) is stable up to 1 180 K, and gamma-phase Fe (FCC) is stable between 1 180 K and about 1 670 K. X-ray diffraction provides the information that the (0 0 1) plane

177

spacing in the alpha-phase is 2.859×10^{-10} m just below 1 180 K, and the (0 0 1) plane spacing in the gamma-phase is 3.592×10^{-10} m just above 1 180 K.

What is the percentage change in volume of a specimen of iron due to the isothermal alpha to gamma-phase transition?

Appendix

PHYSICAL CONSTANTS

Constant	Symbol	Value
Speed of electromagnetic radiation in free space	c	2.9979×10^8 m s^{-1}
Permeability of free space	μ_0	$4\pi \times 10^{-7}$ kg m s^{-2} A^{-2} (or H m^{-1})
Permittivity of free space	ε_0	8.8542×10^{-12} C V^{-1} m^{-1} (or F m^{-1})
Atomic mass unit	a.m.u.	1.6604×10^{-27} kg
Electron mass	m_e	9.1091×10^{-31} kg
Proton mass	m_p	1.6725×10^{-27} kg
Neutron mass	m_n	1.6748×10^{-27} kg
Electron charge	e	1.6021×10^{-19} C
Boltzmann constant	k	1.3805×10^{-23} J K^{-1}
Planck constant	h	6.6256×10^{-34} J s
Avogadro constant	N_A	6.0225×10^{23} mol^{-1}
Rydberg constant (for infinite mass nucleus)	$R_\infty = \dfrac{m_e e^4}{8\varepsilon_0^2 h^3 c}$	1.0974×10^7 m^{-1}
Gravitational constant	G	6.6730×10^{-11} N m^2 kg^{-2}

Appendix

At the following places on Earth's surface the acceleration of gravity, g, has been observed to be:

Africa—Cape Town
   9.796 57 m s^{-2}
Australia—Melbourne
   9.799 87 m s^{-2}
Canada—Ottawa
   9.806 18 m s^{-2}

England—London
   9.811 88 m s^{-2}
New Zealand—Auckland
   9.799 62 m s^{-2}
U.S.A.—New York
   9.802 67 m s^{-2}

Answers

CHAPTER 1

1. (*a*) 62.6 m s^{-1}, (*b*) 1.87 s up, 10.95 s down, (*c*) 44.3 m s^{-1}
2. 1.79 hours
3. $v \propto s^{1/2}F^{1/2}m^{-1/2}$

   or $v = k \sqrt{\left(\dfrac{sF}{m}\right)}$ where k is a dimensionless constant
4. (*a*) 0.20 m s^{-1}, (*b*) 10^{-3} J, (*c*) 2×10^{-2} N
5. 8.33×10^{12} N m^{-2}
6. 1.2×10^{-4} mol m^{-2} s^{-1}
7. (*a*) 2.38 kg m^{-1} s^{-1}, (*b*) 1.89×10^{-3} m^2 s^{-1}
8. (*a*) 8.0×10^{-15} J, (*b*) 50 keV
9. 2.28×10^{-2} m
10. (*a*) 5×10^{-4} m, (*b*) 315 Hz, (*c*) 1.05 m (*d*) 330 m s^{-1}
11. (*a*) 3.14×10^{-2} (or $10^{-2}\pi$) m s^{-1}, (*b*) 9.87×10^3 (or $10^3\pi^2$) m s^{-2}
12. 3.3×10^{-16} J
13. 3×10^{19} Hz
14. 1.4 per cent
15. 7.3×10^{-10} m

CHAPTER 2

1. 3.20×10^{-19} C; 2 electrons
2. 1.608×10^{-19} C
3. 1.76×10^{11} C kg^{-1}

4. $11° 28'$
5. 7.7×10^{-5} T
6. $\lambda_e = 42.8\lambda_P$
7. (*a*) 9.58×10^7 C kg^{-1} (*b*) 1.67×10^{-27} kg
8. (*a*) 1.92×10^5 m s^{-1} (*b*) 2.07×10^{-12} m

CHAPTER 3

1. 0.80 T
2. Mass 16.02—oxygen; mass 32.10—sulphur
3. Mass 63.54—copper; mass 54.94—manganese
4. 28.09
5. 1.875×10^{-6} m
6. 5.292×10^{-11} m; 2.186×10^6 m s^{-1}
7. 1.940×10^{-18} J; 1.025×10^{-7} m
8. 8.716×10^{-18} J (i.e. 54.40 eV)

CHAPTER 4

1. 2.28×10^{-28} kg
2. (*a*) 2.04×10^{-11} J; (*b*) 1.28×10^{-12} J (nucleon)$^{-1}$
3. $Z = 82$; $A = 214$; $Z' = 83$; $A' = 214$; **X** is lead (**Pb**); **Y** is bismuth (**Bi**)
4. (*a*) 0.707 g; (*b*) 2.59×10^{10} s^{-1}
5. Reaction energy: 3.667×10^{-11} J
   (Mass defect: 4.080×10^{-28} kg)
6. 9.397×10^{13} J kg^{-1}
7. 5.33×10^{-13} J (i.e. 3.32 MeV)

CHAPTER 5

1. 9.92×10^{-11} m
2. 1.096×10^{-11} m
3. K_α; 2.788×10^{-16} J; K_β: 2.935×10^{-16} J
4. 1.832 kV
5. 0.125 2 mm
6. 0.706 mm
7. (*a*) 2.09×10^{-2} m; (*b*) 1.045×10^{-1} m
8. 93.28%

CHAPTER 6

1. 0.999×10^{29} m^{-3}
2. 1.18

CHAPTER 7

1. 2.819×10^{-10} m
2. (011): 2.027×10^{-10} m; (012): 1.282×10^{-10} m
3. 5.29×10^{-11} m
4. $5°56'$
5. 7.148×10^4 eV (photon)$^{-1}$
6. 4.53×10^{24} m^{-3}

CHAPTER 9

1. -0.85%

Index

185

Index

Index